Fifty Years of Evolution in Biological Research

Series Editor
Marie-Christine Maurel

Fifty Years of Evolution in Biological Research

Progress and Decline

Jacques Balthazart

iSTE

WILEY

First published 2023 in Great Britain and the United States by ISTE Ltd and John Wiley & Sons, Inc.

ISTE Ltd
27-37 St George's Road
London SW19 4EU
UK

www.iste.co.uk

John Wiley & Sons, Inc.
111 River Street
Hoboken, NJ 07030
USA

www.wiley.com

Library of Congress Control Number: 2023936094

British Library Cataloguing-in-Publication Data
A CIP record for this book is available from the British Library
ISBN 978-1-78630-878-8

Contents

Preface

As I finish writing this book, in the summer of 2022, I am now 73 years old. I am a biology researcher who has been officially retired for eight years, but I continue my scientific activity from a place of passion. As I often told my friends and colleagues, I have been paid my entire working life for an activity that was also a hobby, and I see no reason to stop this activity just because my salary has been turned into a pension. This being said, this "job", if it is a job at all, has changed profoundly over the course of my 50-year career, sometimes for the better, but often for the worse. This is what I would like to explain in this book while evoking what the life of a researcher in the fundamental sciences interested in knowledge can be, independently of the profit that can be made from it.

I initially experienced the profound changes in research conditions during my career with enthusiasm. This was probably related to my naivety and inexperience, but retrospectively, I really think that most of the changes that occurred in the research world during the last quarter of the 20th century have had largely positive effects. First of all, there have been significant changes in analytical techniques, particularly at the biochemical and then at the genetic level, which have made it possible to tackle questions that were previously impossible to study. In addition, research tools have become considerably more sophisticated. First, calculation tools and also all the equipment used in the laboratory have become miniaturized and automated. These changes have rapidly made it possible to carry out a much larger amount of work in one day.

However, at the same time, new constraints and obligations have gradually developed, while the expectations of what the researcher should produce have grown exponentially. This initially led to a limitation of the researcher's ability to make progress, but then we saw a gradual decrease in the quality of research, for some in any case, as a result of the increasing pressure to publish that gradually developed.

It is my gradual discovery of this state of affairs that has prompted me to pick up my pen and try, first of all, to take stock of the current situation, to identify as precisely as possible the causes of the problem and to propose the solutions that seem possible at this stage. I hope that this short book will serve as a catalyst for the scientific community and the organizations that fund it to correct this state of affairs as much as possible. That said, research has by now become a commercial enterprise not quite like any other, but is nevertheless subject to significant constraints. Even so, I hope that this creative activity at the highest level and which is eminently satisfying can survive in a disinterested form by promoting above all curiosity and the thirst for learning.

May 2023

Acknowledgments

In retrospect, I think I was truly blessed to have lived a life as a research scientist, both for all the happy moments I experienced and for what was accomplished scientifically. This of course would have been impossible without the help, mentoring and collaboration of so many people that it is clearly impossible to list them here. This list includes:

– the exceptional teachers I had the honor of having during my primary and secondary education – I think especially of my French, physics and biology teachers who gave me the basics to learn science effectively;

– my professor of zoology at university, Michel Chardon, who pushed me to switch from veterinary studies, which I had started by mistake, to zoology, a step I have never regretted;

– my PhD supervisor, Ernest Schoffeniels who gave me total freedom to develop the research project I had in mind even though this project was quite far from the interests of his laboratory. Nevertheless, he was always completely supportive of my research and a source of endless enthusiasm.

All the work I have published has been carried out with a large number of collaborators, including more than 30 students doing their undergraduate thesis (now renamed graduate work), 21 PhD students, 17 postdoctoral students, 15 international collaborators who stayed in the laboratory for long periods of time and finally my long-time collaborators with whom I have worked for 10, 20 and sometimes 30 years with great pleasure and success.

It is impossible to mention all these people here and I will only mention the collaborators who have played a major role in this story:

– Michael Schumacher, who developed the first measurement of aromatase activity in the laboratory.

– Michelle Baillien, who developed a faster test for the activity of this enzyme, based on the measurement of tritiated water production. She was also instrumental in launching our research program on rapid regulation of brain aromatase activity.

– Gian Carlo Panzica and his wife Carla Viglietti, from the University of Turin, who taught me neuroanatomy and with whom I have published many articles since 1986.

– Annemie Van der Linden (University of Antwerp), who introduced me to the world of in vivo nuclear magnetic resonance imaging. We have published regularly together since 1998.

– And of course Gregory (Greg) Ball, my scientific twin. He developed an interest in ethology, endocrinology and neurosciences through paths very similar to mine. Having received very similar training, we understood each other from the beginning implicitly. In fact, one word evoked a whole stream of ideas for us where others would have needed whole sentences to understand. This collaboration which began in 1985 can be considered a model, I think, of creativity, productivity and, above all, fun!

I would also like to offer special thanks to Charlotte Cornil who has worked in collusion with me since her dissertation in 1999 and took over the leadership of the behavioral neuroendocrinology research group when I became emeritus in 2014.

And finally, I would like to publicly thank my wife Claire. She has not only given me 50 years of love and two beautiful boys who now have children of their own, that is our private life, but more relevantly for this discussion concerning science, she has also been my friend and companion during all these years, attentively encouraging me when I was demoralized, giving me advice when I was in doubt and trying to calm me down when I was irritated. And, above all, she created for me an incredibly favorable environment for me to develop my work and my passion. For many, many years, she took on most, I should probably say almost all, of the burden of daily life: all the tasks that go into raising a family and maintaining a functional and very pleasant home. I will be forever grateful to her for what she did for all those years when I was developing my scientific career. Thank you, Claire, I love you!

Introduction

This book speaks of the profound changes which have affected the conditions in which fundamental scientific research is pursued, i.e. research motivated essentially, if not exclusively, by curiosity and the desire to discover the laws that govern the functioning of the universe. It concerns very directly research in biology and more particularly the field of behavioral endocrinology which was the subject of the author's studies, but most of the observations and conclusions apply *mutatis mutandis* to all fields of fundamental research.

Fundamental research changed drastically during the second half of the 20th century and the beginning of the 21st century. In 1950–1960, researchers were still working largely independently and collaborations were still a rarity. Funding for laboratories was plentiful and not closely linked to their performance. The university where I worked from 1971 to the present had large budgets directly from the government and redistributed these funds to the various laboratories that could easily conduct their research activity without spending considerable time writing research proposals to obtain their funding. This situation changed in the 1970s when laws regulating the funding of universities in Belgium drastically reduced it. A similar change also took place in other Western countries. This progressively initiated a race for grants and in parallel a race for publications, which became the only criterion for the quality of research. At the same time, from 1975 onwards, a quantitative method of evaluating the so-called quality of scientific journals was developed, the impact factor, which measures in a comparative way the quantity of citations of articles published by a given journal within the past two years. Quite quickly, the impact factor was more or less unconsciously transformed into a measure of the quality of the articles themselves and thus of the work of the researchers. This has resulted in a race to publish in journals with a high impact factor, which in my opinion has been very damaging for the overall quality of research. I will come back to this in more detail.

At the same time, and whether or not this is related, we can discuss it, we have seen a significant decrease in the confidence that the general public has in the scientific process over the last 50 years. In 1960–1970, the general public thought that science was able to solve all our problems and to lead us to the Moon, which it did. Paradoxically, while scientific progress has never been so important, we have witnessed in the last few decades a resurgence of various irrational beliefs and a questioning of the value of the scientific consensus. I will try to explain, at least in part, this surprising shift in public opinion. The "profession" of researcher in biology has thus changed considerably over the last 50 years, for the better (technical progress) and also in a less positive way (various constraints, relative disaffection of the general public). It is this change that I would like to present here.

This description is largely based on the author's personal experience, who for more than 50 years has been actively involved in various research activities in the field of biology and more specifically in the field of behavioral endocrinology. This activity has led to two fundamental observations. On the one hand, it has led to a clear awareness of the fact that changes in research methods and conditions have a profound effect on the career of a particular individual. On the other hand, it has become increasingly clear that fundamental research cannot be planned. Chance, more elegantly called serendipity, plays a very important role in its evolution. We will come back to this when we discuss research funding methods.

The author's scientific activity has indeed shifted over time from one issue to another. Even though all these questions were intellectually linked, it is clear that serendipity has played an important role in this evolution. These changes in research themes were made through discoveries, meetings at conferences, contacts with other researchers first by mail and then by e-mail, and finally through the reading of particularly impressive articles. This accumulated experience shows that it is particularly vain to try to plan fundamental research. Discoveries are made by chance meetings between different ideas, between different researchers or following the appearance of a technique that allows a new question or an old question to be approached from a new angle.

Over the past few decades, there has been an increasing tendency to allocate research funds to respond to specific questions defined by planning bodies. Thus, the majority of research credits distributed by the European Union finance projects submitted in response to specific requests in a defined field. This is what is known as the "top-down" approach: researchers are asked to solve a specific problem. This approach works, of course, in applied research, but not in fundamental research. For a long time, fundamental research was based on a "bottom-up" approach, where researchers propose a work program based on their interests and skills. The federal research support agencies in the United States, the NIH (National Institutes of Health) and the NSF (National Science Foundation) have made extensive use of this funding method. This funding mode was also widespread during the second half of the 20th century among the national funds that finance research in the countries of Western Europe (Belgium, France, Italy, etc.). Many countries, however, have gradually transferred their resources to the European Community, which has become a major source of research funding in Europe and has operated for 20–30 years almost exclusively using the "top-down" model. This attitude has been problematic for fundamental research and has led to the virtual disappearance of entire fields of research in Europe. The creation of the ERC (European Research Council) in 2007 has fortunately partially reversed this situation, and this organization now funds research programs initiated by the researchers themselves. It is only regrettable that the amount of money distributed by the ERC remains far less (less than 20% of the total) than that made available to researchers via the interventionist approach. One of the goals of this book will therefore be to try to convince as wide an audience as possible that the planning of fundamental science leads largely to its sterilization.

1

The Evolution of Techniques

In the last few decades, the techniques for investigating biological phenomena have developed in a literally extraordinary way. This was probably the "golden period" of biology, following the rapid development of chemistry and physics in the previous two centuries. It may be useful to briefly summarize these technical developments here insofar as they are relevant to the author's field of research, namely behavioral neuroendocrinology. They have also conditioned the evolution of research in general which is the subject of this book. This evolution has recently been presented with more technical detail in a special volume of the journal *Hormones and Behavior* published on the occasion of its 50th anniversary (Balthazart 2020). Eight main types of technical developments should be mentioned.

1.1. Hormone assays

Increasingly sensitive techniques have been progressively developed, enabling hormones present in small biological samples to be determined. In the 1960s, circulating hormones could only be measured by biological assays which, although specific, were not very sensitive. Thus, prolactin, a pituitary hormone, was measured in samples by injecting them into pigeons or ring doves and measuring a few days later the growth of the crop (a bulge in the digestive tract that secretes a caseous liquid, the crop milk used to feed the young), which is directly controlled by this hormone (Nicoll 1967). In the same way, a pregnancy test in females was achieved by injecting urine extract into rabbits and observing whether this induced ovulation (Cabrera 1969). Studies including hormone assays were therefore very rare. The first volume of the journal *Hormones and Behavior* published in 1969 contained only two articles including such assays performed in large volumes of urine, whereas such assays are now the rule in most publications.

Chemical methods (fluorimetric assays, paper chromatography, spectrophotometry, gas chromatography) were gradually developed (Jallageas and Attal 1968), but these time-consuming and laborious methods were only slightly more sensitive and only allowed the determination of hormones in large volumes of blood from large animals. In 1959, Solomon Aaron Berson and Rosalyn Yalow developed a first radioimmunoassay for measuring insulin in small volumes of blood (Yalow and Berson 1959). This discovery, which earned the authors the Nobel Prize in 1977, marked the beginning of a revolution in biological and biomedical research, even though this new technique was initially greeted with skepticism. The technique was quickly applied to the determination of protein hormones, but it took another 10 years before this type of determination could be adapted for steroid hormones such as estradiol and testosterone (Abraham 1969; Furuyama et al. 1970).

The exquisite sensitivity of the radioimmunoassay which goes down in the best cases to a few picograms (10^{-12} g or one thousandth of a billionth of a gram) has made it possible to measure these hormones in a few microliters of blood or in microbrain samples. This technique is also relatively fast and simple, allowing hundreds of samples to be measured quickly. This made it possible for the first time to undertake a long-term study of the evolution of the hormonal environment of small species such as mice or songbirds (see, for example, Romero and Wingfield (2016) and Wingfield (2017)).

More recently, a new approach to steroid assays has been developed. It is based on a fine separation of the different steroids present in a sample by chromatography or high performance liquid chromatography (HPLC) followed by their identification and quantification by mass spectrometry (Liere et al. 2000; Jalabert et al. 2021). This approach can determine dozens of different steroids in the same very small sample size (sensitivity equal to or better than radioimmunoassay) with an unequalled degree of specificity. The only disadvantage is that only one sample can be assayed at a time, even though the analysis can be automated in sequence for multiple samples. It also requires the use of sophisticated and expensive equipment. The combined use of these two techniques, however, provides unparalleled opportunities for the analysis of relationships between hormones and physiological processes including behavior.

1.2. Techniques for the identification of steroid hormone action sites in the brain

In the early 1970s, the sites in the brain where sex steroids (testosterone, estradiol, progesterone) act to activate sexual behavior and other physiological mechanisms were essentially unknown. The appearance at that time of radioactive steroids made it possible for the first time to address this issue effectively. Two laboratories led by Donald Pfaff at the Rockefeller University in New York and Walter Stumpf at the University of North Carolina in Chapel Hill had independently developed the in vivo autoradiography technique for visualizing these binding sites (Pfaff 1968a, 1968b; Stumpf 1970; Sar and Stumpf 1972). These sites are mainly concentrated in the preoptic area, the hypothalamus and some nuclei of the telencephalon such as the amygdala and the bed nucleus of the stria terminalis. They represent in fact the places where nerve cells express receptors for these steroids, namely, androgen, estrogen and progesterone receptors (Kelley and Pfaff 1978; Morrell and Pfaff 1978). The mapping of these sites could, over the next two decades, be confirmed by visualizing corresponding receptor proteins or RNA messengers respectively by immunocytochemistry or in situ hybridization. Over time, the sensitivity and specificity of all these methods have progressively improved, and they have thus made it possible to localize with cellular resolution all the sites in the brain where sex steroids act.

For the researcher interested in the control mechanisms of behavior, however, the identification of these binding sites was only a first step in the investigations. It remained to establish which of these sites are directly involved in the activation of behavior. This goal was progressively achieved from 1960 onwards thanks to the development of stereotactic surgery techniques which make it possible to modulate in an anatomically specific way the activity of precise nerve sites (Lisk 1960; Michael 1962; Barfield 1969). To do this, the anesthetized animal is fixed in a standardized way in a metal frame which carries gallows calibrated to a tenth of a millimeter. These then make it possible to guide either electrodes in the brain to produce microlesions in sexually active animals or cannulas which will make it possible to inject or implant steroids in animals deprived of their gonads and therefore sexually inactive. It has thus been possible to establish in a variety of animal models (mammals, birds, reptiles, etc.) that the sexual behavior of the male is largely controlled by the action of androgens or their estrogenic metabolites which are derived by aromatization mainly in the preoptic area. The sexual behavior of the female, on the contrary, depends on the action of estrogens, possibly associated with progesterone, at the level of the hypothalamus, especially the ventromedial hypothalamus. This being said, it is clear that many other brain sites are involved in the control of these behaviors, but their description is clearly beyond the scope of this book. The interested reader can find a synthesis of this research in more specialized journal articles (Hull and Dominguez 2015; Pfaus et al. 2015).

1.3. Molecular biology and sequencing techniques

The last 50 years have also seen the rapid development of incredibly sophisticated and increasingly efficient techniques for isolating and characterizing nucleic acids, DNA (deoxyribonucleic acid) and RNA (ribonucleic acid) which are the basis of the genetic code and its use in all living organisms.

The chemical structure of DNA was formally identified in 1953 by James Watson and Francis Crick (Nobel Prize in Physiology or Medicine in 1962) as essentially consisting of a double chain formed in a specific order of four different components, the nucleotides adenine, thymine, cytosine and guanine (abbreviated as A, T, C, G) (Watson and Crick 1953). However, it was not until 1977 that a reasonably easy method for determining the sequence of this DNA appeared, the so-called Sanger method, which also earned its discoverer the Nobel Prize in 1980 (Sanger et al. 1977). However, this method could only determine the sequence of short portions of DNA (a few hundred nucleotides) and the sequencing of a long DNA chain therefore required the analysis of multiple partially overlapping segments, which was time-consuming and laborious. More sophisticated and rapid methods have developed over the years, however, and these advances still continue today (Heather and Chain 2016; Shendure et al. 2017). As a result, the speed of DNA sequencing has improved dramatically and its cost has decreased proportionately. As an example, the first complete sequencing of the human genome published in 2000 cost between $300 million and $1,000 million and involved hundreds of researchers over a period of around 13 years

(Lander et al. 2001; Venter et al. 2001). The same work can now be done in a day for about $1,000! And there is reportedly a new sequencing technique that puts the human genome at about $100 (Pennisi 2022).

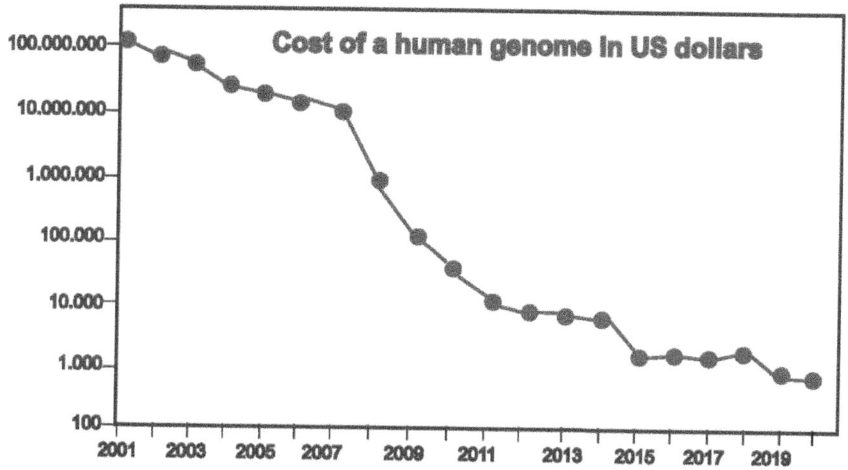

Figure 1.1. *The phenomenal decrease in the cost of DNA sequencing illustrated by the cost of sequencing a human genome[1]. For a color version of this figure, see www.iste.co.uk/balthazart/fiftyyears.zip*

We have thus moved from the first-generation Sanger sequencing technique, to the second-generation techniques known as Illumina methods, and then to the third-generation methods or the Nanopore method[2]. Illumina sequencing techniques use reversible chain-terminating dyes to identify DNA sequences. The sample is first cleaved into short sequences that are then bound to generic adapters, and each fragment is amplified into many copies by the polymerase chain reaction or PCR technique (see end of this section). Each fragment is separated into single strands and sequenced in a medium that contains the different fluorescently labeled nucleotides, DNA polymerase, and a terminator that interrupts the PCR cycle. As a result of the terminator, only one base is added per strand to each cycle. The cycle terminator is then removed allowing the addition of the next fluorescent base. Based on the fluorescent signals emitted, a computer detects the base

1 Redrawn from data in: www.genome.gov/about-genomics/fact-sheets/Sequencing-Human-Genome-cost.

2 See www.differencebetween.com/difference-between-nanopore-and-illumina-sequencing.

that has been added in each cycle. This approach makes it possible to identify a DNA sequence in 4–56 hours, depending on the length of the chain in question. However, this approach only allows the sequencing of short strands of 100–150 base pairs, which must then be reassembled in a rather tedious process.

The Nanopore technique is based on an entirely different principle in which the passage of DNA through a very small pore (nanopore) sequentially induces a signal characteristic of each nucleotide which can therefore be recorded. In general, a flow cell has a number of nanopores that are integrated into an electro-resistive cell. Each nanopore is associated with an electrode and a sensor chip that measures in real time the current flowing through the nanopore. This current is characteristic for each nucleotide, which makes it possible to determine the sequence of the nucleic acid in real time without the need to use modified nucleotides or to split the DNA into short sequences. This process is therefore much less laborious, less expensive and much faster.

This technical progress has opened the door to scientific studies that, until recently, would have been considered science fiction. The genome of a rapidly growing number of species is now known and provides valuable information on the evolution and its mechanisms. One example is a series of studies that have sequenced the genomes of 48 species of birds representing the different major existing groups. These data have led to a better understanding of the evolution and adaptation of these birds to different lifestyles (Zhang et al. 2014), a better understanding of the evolution of their beaks from a tooth-bearing jaw (Meredith et al. 2014) or the acquisition of learned vocalizations (Pfenning et al. 2014). Subtle genetic mutations and differences between individuals are being identified and can be linked to various pathologies which allows for considerable medical progress and will lead in the near future to the development of personalized medicine. Ancient DNA can be detected and sequenced, which sheds new light on the evolution of the human species (Paabo 2015). And these are just a few of the countless examples to which these new techniques are being applied.

Multiple techniques have also been developed to analyze and quantify the expression of genes contained in this DNA. The RNA resulting from the expression of DNA (transcription) can be easily identified, amplified and quantified in a relatively simple manner, thanks in particular to the polymerase chain reaction or PCR, technique (Mullis et al. 1986) whose name has become familiar to the general public since the Covid-19 pandemic. The set of genes expressed at a given time in an organism, an organ or even a single cell can thus be determined quantitatively and linked to the physiology of the structure. This technical approach enables an analysis of the physiology and behavior of an individual, whether healthy or suffering from a pathology, with an unprecedented precision. This leads to considerable progress in both fundamental biology and medical research.

1.4. Controlled modification of gene expression

Advances in molecular biology have naturally made it possible to specifically study the physiological role of genes identified by sequencing whose transcription, and therefore, the concentration of the corresponding RNA) is correlated with the expression of a specific behavior. This approach has, of course, also been developed in all other fields of biology and medicine. A technique was first developed that allowed the expression of a gene and thus of the corresponding protein to be completely inactivated in mice (Capecchi 1989). This gave rise to numerous knock-out (KO) mouse lines which made it possible to specify the physiological role of a given protein, for example, the estradiol receptor (Lubahn et al. 1993; Krege et al. 1998) or aromatase, the enzyme that controls the conversion of testosterone to estradiol (Fisher et al. 1998; Honda et al. 1998).

By 1980, this technique fundamentally transformed the ability of researchers to analyze the function of individual genes in mice. The National Institutes of Health (NIH) in the United States actually allocated hundreds of millions of dollars over the next two decades to generate KO mice for numerous genes resulting in the production of more than 4,000 different lines. They then realized that it was necessary to coordinate all of these efforts which led to the Knockout Mouse Project (KMOP) Repository (the archive of KO mice) which stores and distributes live mice that are KO for more than 8,500 different genes and also the corresponding engineered stem cells and frozen embryos (Lloyd 2011).

However, this method suffered from a number of practical and conceptual problems. The comparison of KO mice and control mice allows us to define functional deficits linked to the absence of expression of a specific protein, but it is often impossible to confirm the conclusions obtained by the reverse experiment, which would consist of restoring the affected function.

An exception to this problem is when the KO affects an enzyme such as aromatase. It is then possible to restore the function affected by suppressing the transformation of testosterone into estradiol by administering exogenous estradiol to the affected animal (Bakker et al. 2004).

Moreover, for practical reasons that are only partially understood, the KO technique has so far only been applied extensively to mice of a particular strain (Sv129), which limits the possibility of generalizing the results obtained, especially because these mice have a large number of peculiarities such as the absence of the corpus callosum, the main communication between the two cerebral hemispheres (Wahlsten 1982). The recent development of the technique known as CRISPR-Cas9, commonly known as *molecular scissors*, which allows for easy modification of a specific point in an individual's genome (by Doudna and Charpentier (2014), Nobel Prize in Chemistry 2020) has removed this limitation by potentially creating KOs in all animal species (Doudna and Charpentier 2014).

Finally, and more embarrassingly, suppression of gene expression by KO is present from early development and throughout the life of the individual. These early deficits often induce compensatory mechanisms that mask functional deficits (see, for example, Apostolakis et al. (2002)). Moreover, it is impossible to separate effects related to developmental changes from those that specifically affect function in adults. The solution to the latter problem, however, has been found by developing conditional KOs in which the DNA of a gene has been altered so as to suppress the expression of that gene, but the suppression becomes effective only if the animal is exposed to a chemical trigger (e.g. injection of an antibiotic or tamoxifen, an anti-estrogenic compound not normally present in animals). This leads to having normally developing subjects in whom suppression of the protein of choice is induced at the desired time (Sauer and Henderson 1988; Zhang et al. 2012). More sophisticated variations of this method have also been developed that allow, for example, the expression of a protein to be suppressed uniquely in a particular cell type, such as only those cells that express a given receptor or a given neurotransmitter (Yeh et al. 2002; Gibson and Muse 2009; Raskin et al. 2009; Naule et al. 2016). However, these approaches are not without methodological problems and require the performance of controlled manipulations that are unfortunately not always implemented (Harno et al. 2013).

1.5. Techniques for temporarily modifying the activity of neurons

In all neurosciences, and in particular in behavioral neuroendocrinology, technical progress has made it possible to manipulate the activity of neurons or groups of neurons in an increasingly precise manner in order to determine their specific role. The stereotaxic surgery techniques already mentioned make it possible to damage or administer a hormone in an anatomically precise manner to a limited group of neurons. This approach was initially used to delineate groups of cells or nuclei involved in the activation of specific behaviors. It was then rapidly combined with the use of pharmacological tools that, for example, stimulate or block specific steroid receptors or inhibit the transformation of one steroid into another, for example the aromatization of testosterone into estradiol.

Molecular biology has also come to the rescue in this endeavor by making it possible to temporarily block the synthesis of proteins suspected of being involved in the control of a physiological function or behavior. A variety of technologies based on blocking the translation of a specific RNA into the corresponding protein have been developed (McCarthy et al. 1993a). Any protein results from the transcription of a gene present in the DNA into an RNA that will itself be translated into a protein. It is possible to block or at least greatly reduce this translation by injecting an RNA fragment whose sequence is complementary to that of the targeted RNA. This injected RNA is called antisense and, by binding with the target RNA, it will prevent its translation and promote its degradation, so that the production of the corresponding protein will be temporarily inhibited until a new stock of this RNA is transcribed from the DNA. It will thus be possible to evaluate the function of the protein concerned following this blockage obtained by the injection of antisense or similar compounds (morpholinos, small inhibitory RNA or siRNA, etc.). This technique has namely been used to assess the role of the estrogen receptor in the masculinization of rodent behavior (McCarthy et al. 1993b) or of the steroid receptor coactivator SRC1 in the activation of sexual behavior in female rats (Molenda et al. 2002) or male quail (Charlier et al. 2005).

Several other techniques have also been developed in recent years to modulate the activity of specific neurons in an even more specific way. One of them involves inducing in these cells, by genetic manipulation (e.g. via CRISPR-Cas9) or by injecting a virus containing the gene of interest, the expression of a receptor that is only activated in the presence of a compound that does not exist physiologically in the body. This type of receptor is called DREADD (Designer Receptor Exclusively Activated by Designer Drugs), a receptor created to be activated exclusively by a synthetic drug. This receptor, which belongs to a very specific class of membrane receptors, the GPCRs (or G protein coupled receptors), is then incorporated into the membrane of the target cells (the entire brain or a specific class of neurons that have been targeted by the virus), and in the presence of the synthetic drug (very often CNO or clozapine N-oxide) it will depolarize and induce the electrical activity of that cell. This DREADD technique therefore makes it possible to test the role of a specific group of neurons by rapidly inducing (a few seconds after the injection) for a relatively short time (a few minutes to an hour) their activity via injection of CNO.

An even more timely activation of a few seconds of specific groups of neurons can finally be obtained by a technique known as optogenetics, which combines light stimulation with a genetic manipulation of the content of these neurons (Boyden et al. 2005; Deisseroth et al. 2006). In this technique, the expression of a light-sensitive membrane ion channel is induced in the target neurons, by genetic manipulation or viral injection. When exposed to a specific wavelength of light, this channel can, depending on its nature, either depolarize (activate) or hyper polarize (inhibit) the cell that contains it with a temporal resolution of a few milliseconds. To study the role of these neurons in the control of a behavior, it is then possible to implant an optical fiber precisely in the area of the brain located dorsally to the site that expresses the light-sensitive channel. A point illumination with light of a precise wavelength emitted at high frequency (2–20 Hertz) by a light source connected to a computer will then make it possible to instantly activate or block the activity of the neurons expressing the light-sensitive channel and thus to demonstrate their role in the control of a given behavior or function. This approach now allows a direct causal analysis of the role of specific neural structures in the control of behavior, a task that seemed completely impossible only a few decades ago.

Over the last 50 years, these technical developments have had a profound influence on our knowledge of behavioral neuroendocrinology, biology and medicine. In neuroendocrinology, these developments concern in particular the mechanisms that control sexual differentiation of behavior (e.g. the discovery of "direct" effects of genes on sexual differences, independently of the effects of sex steroids (Arnold 2017) and epigenetic effects (McCarthy and Nugent 2013; McCarthy et al. 2017)). They have also led to a better understanding of behavior activation by steroids in adulthood (e.g. the existence of steroid synthesis in the brain, a more complete understanding of the interaction of steroids with their receptors and the role of coregulators, identification of the rapid effects of steroids initiated at the level of the cellular membrane, identification of the molecular mechanisms underlying neuronal plasticity, etc.). A description of these conceptual evolutions is beyond the scope of this book but can be found in a recent journal article (Balthazart 2019).

1.6. The rise of the computer and the personal computer

Many of these technical developments would have been impossible if there had not been at the same time an incredible development of computing and data processing tools. It is difficult to imagine today that 50 years ago there were no personal computing tools. Personal computers and even hand-held calculating machines did not exist at that time. All calculations had to be done either mentally or with the help of a slide rule, a mechanical instrument that enabled multiplication and division by simply moving a graduated slide lengthwise (Figure 1.2). For any more complicated operation, the only solution was to use a large mainframe computer, present in universities or large companies, with which one communicated by means of punched cards. At that time, such a computer occupied tens of square meters for a very low computing power, much lower than that of a current smartphone. The personal anecdote presented in Box 1.1 illustrates, I think, this limited computing power.

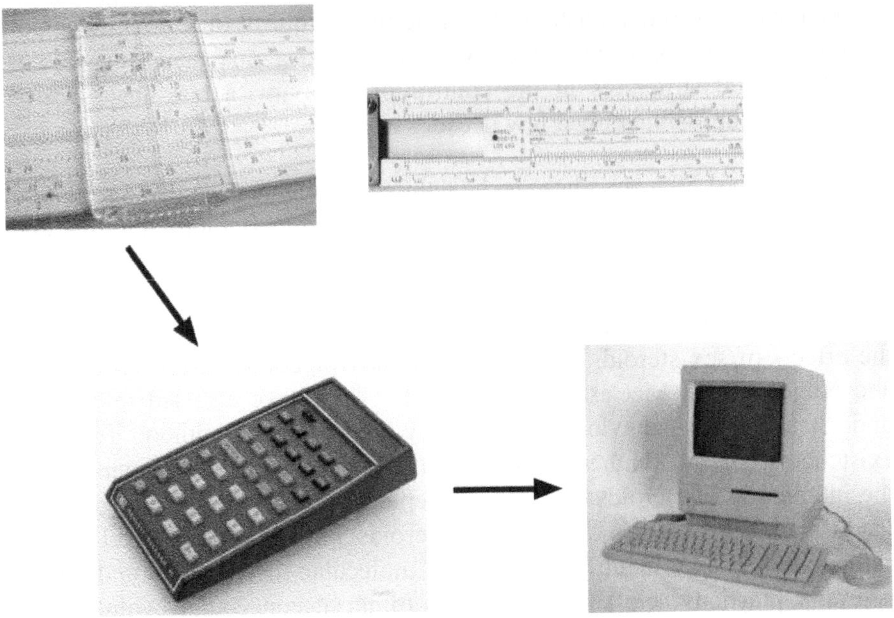

Figure 1.2. *Evolution of calculating instruments, from the slide rule, the only way to do simple calculations 50 years ago, to the first personal computers, via the first calculators. For a color version of this figure, see www.iste.co.uk/balthazart/fiftyyears.zip*

This situation began to change rapidly during the 1970s. The technical means available for research were clearly progressing rapidly, but they would still seem very primitive to today's researchers. For example, while radioactive isotopes were readily available and could be widely used for radioimmunological assays, the results could only be calculated with rudimentary means. Initially, these were the slide rule already mentioned or a so-called electric calculating machine (an Olivetti Divisuma 24, Figure 1.3), but which was in fact only a mechanical machine whose cogs were driven by small motors. The machine was designed to add, subtract, multiply and divide "at high speed" in any order. Each step and the result of the calculation was printed on a roll of paper (15 columns, including 13 numbers and two signs). The division was, according to the producer, completely automatic, but took from 3 to 10 seconds depending on the number of transfers that had to be made from one column of digits to the next. Moreover, the mechanism did not generate any decimal automatically and the division of 10 by 3 gave 3 as result. To obtain more precision, it was necessary to add decimals manually (e.g. 10.00/3 = 3.33). The calculation of a series of 100 RIA determinations took several hours.

For my graduate work (1971), I decided to compare the correlations obtained by several different calculation methods, including a non-parametric method, the Spearman correlation coefficient. Dr Gilbert Laplanche from the computer center at the University of Liege in Belgium, had written the program allowing all these calculations on the available IBM computer. The calculation of the Spearman coefficients in question first involved arranging series of about 150 numbers in ascending order. The computer was accessed by dropping in a predetermined location a packet of punched cards containing the program and the data to be analyzed. The access request was made under certain conditions: one asked either 30 seconds of calculation time, or 10 minutes, or one hour or even more and that conditioned the moment when the punched cards would be loaded in the computer, either respectively in the next 30 minutes, 2 hours, 6 hours or during the following night. All our first attempts at calculation with deposit in the 30 seconds category were unsuccessful. Gilbert Laplanche thought for a long time that there was a programming error, but numerous checks did not identify it. He therefore progressively submitted this program in the 10-minute, one-hour and unlimited duration categories, and it was only in this last option that the results were finally obtained. The classification of the series of 150 data required about an hour and a half of calculation in this IBM. The 2015 iMac on which I am writing this text now does this in a second!

The technology was far from what it is today. This affected all aspects of research, including the production of written reports. Any work to be produced in multiple copies had to be typed graphically by a professional secretary on alcohol stencils which then allowed a number of copies to be made. All graphics used to illustrate the text were drawn in India ink on tracing paper by a professional and then photographed in black and white, and finally, the photos were printed and then pasted separately into each copy of the document. The preparation of a written work was thus something tedious and expensive! We will come back to this.

Box 1.1. *Anecdote: the state of computing in 1971*

The appearance of the first electronic calculators at the beginning of the 1970s, which became widespread in the following years, was therefore a real blessing. But, one of the first Hewlett Packard (HP) calculators that appeared at that time and that performed instantly (!) the four operations and the trigonometric calculations unfortunately cost more than 50,000 Belgian francs at that time (€1,250 without taking into account the devaluation of money by a factor equal to or greater than 10 since the mid-1970s). So the relative price of things has changed!

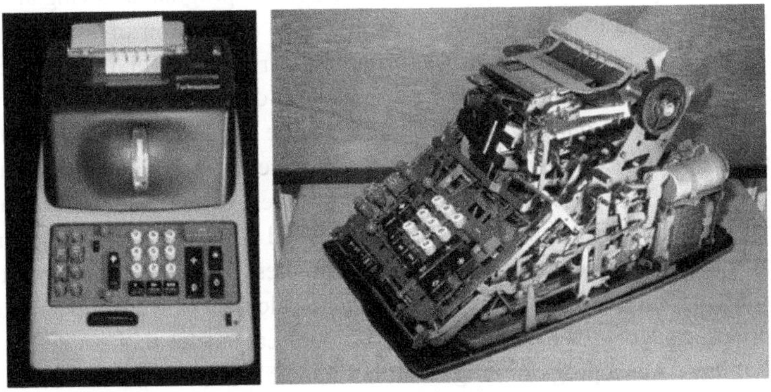

a) b)

Figure 1.3. *(a) The so-called electric calculator, in fact mechanical, Olivetti Divisuma 24. (b) The cover has been removed to reveal the underlying mechanics*[3]*. For a color version of this figure, see www.iste.co.uk/balthazart/fiftyyears.zip*

The possibilities of calculation then progressed in a major way. This led to the introduction of large, programmable table calculators that could, for example, calculate the mean and standard deviation of a group of data in one go, a task that had previously been very difficult. These results then had to be re-entered to compare the two means with a Student's test, because the magnetic cards used for programming could not contain enough program steps to calculate everything at once. At that time, such a machine cost more than 200,000 Belgian francs (€5,000 without taking into account the devaluation of the currency). The price of this type of equipment then

3 https://en.wikipedia.org/wiki/Calculator.

dropped dramatically and disappeared in favor of pocket calculators that had become programmable or personal computers.

Reasonably, efficient personal computers, however, only gradually emerged in the 1980s. At the end of the 1970s, the analysis of a large amount of data could still only be done on a large IBM-type computer available only in universities or companies. These analyses required typing ("punching") the data to be analyzed on punched cards, depositing the calculation requests in separate packets, recovering the results, often correcting typing errors that induced a rejection of the procedure by the computer (no error was, of course, allowed) and finally synthesizing the results.

Communication with mainframe computers has also been progressively improved and has made it possible to consider changes in the way scientific papers and, more broadly, all published texts are written. The Institute of Animal Behavior at Rutgers University (Newark NJ, USA) where I was doing postdoctoral studies had in 1978 installed an embryonic form of word processing based on a large computer located on a campus more than 40 kilometers away in New Brunswick. The lab had a teletypewriter that was actually a kind of typewriter called Teletype that could be connected via telephone to a mainframe computer located on the New Brunswick campus. The connection was made simply by calling a number through the telephone and inserting the speaker and microphone phone of the horn into two suction cups associated with an MODEM (modulator-demodulator). Each keystroke on the keyboard produced a specific sound in the loudspeaker, sound that was transmitted to the computer that converted it into a character. All the information, the text entered, was thus stored in the computer and could be kept there. It was then possible to correct the text, modify it, lay it out, and finally recover it to print it on the Teletype. We could thus rework a text without having to rewrite everything and type the new text on the machine. The system had its limitations, however, including the lack of automatic data storage: if the phone line cut off abruptly, which sometimes happened, text that had not been specifically saved was lost.

These were the beginnings of word processing on computers, which would soon become widespread and possible on small computers that every researcher would soon have on his/her desk. Thus, the firm Radio Shack (Tandy) produced in 1977 a first personal computer, the TRS-80 (Tandy Radio-Shack) model I which unfortunately cost the tidy sum of approximately 100,000 Belgian francs (€2,500 without counting the 40 years of devaluation). However, this machine was incredibly primitive by today's standards. It had only 4K of RAM (Random Access Memory) and 1K of video memory which only allowed monochrome display on 64 columns and 16 lines of characters which could only be numbers and capital letters. Saving of the data was carried out on magnetic tapes using a reader/recorder of musical audio cassettes. This saving and recovery was therefore very slow and associated with frequent data loss.

This model I was followed two years later by a TRS-80 model III with a basic memory of 16 K, expandable to 48 K. Importantly, it had two floppy disk drives (5.25"), which made it possible to save data in a much safer way. Each floppy disk could initially hold 360 K, a value that was soon increased to 1,200 K. For the first time, a researcher had on his desk a personal computer that allowed him/her to do word processing, to carry out relatively sophisticated statistical analyses (Student's tests, analyses of variance, correlations, etc.) and to draw scientific graphs (histograms, regression lines, etc.) provided of course that he/she wrote the corresponding programs himself/herself in BASIC language; the existing programs were rare and largely limited to text processing. This meant that the researcher had to spend many hours learning a programming language (BASIC or FORTRAN, which were the most popular at the time) and then writing the programs he/she needed.

Over the past 40 years, these personal computers have continued to evolve rapidly. The TRS-80s were replaced by Apple IIEs and then by the Macintosh line from the first 128K Mac launched in 1984 to the current iMac and MacBook Pro or MacBook Air laptops[4]. Competing companies, primarily IBM, also entered this market at this time and developed alternative offerings, although the Macintoshes remained very broadly established in academic circles.

4 https://en.wikipedia.org/wiki/Macintosh_128K.

For a sum of between €500 and €2,000, any researcher can now have on his desk a machine with a wide range of possibilities: word processing, any form of calculation, drawing, image or sound analysis. A RAM of 8–16 gigabytes (Gb) is now a basic value, and the storage capacities on hard disks, which have long since replaced floppy disks, are currently equal to 512 Gb or 1 or even 2 terabytes (Tb). Moreover, these machines are almost always connected to the Internet, which provides easy access, on the one hand, to huge masses of information and, on the other hand, to enormous computing possibilities via connection to clusters (see section 1.7).

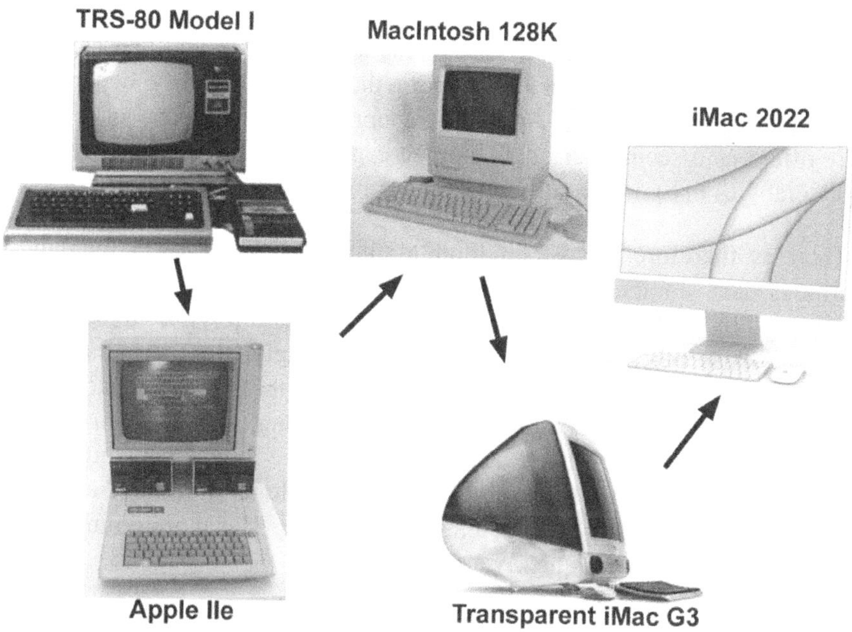

Figure 1.4. *Forty years of evolution of personal computers. For a color version of this figure, see www.iste.co.uk/balthazart/fiftyyears.zip*

The "big" computers in the computing centers have also experienced a spectacular development. Fifty years ago, my university's general computer service had a lamp machine occupying dozens of square meters with very modest power, an IBM series 360 with a central memory of a few dozen kilobytes (KB). It now has multiple Unix/Linux type computer systems associated in several clusters dedicated to intensive computing. One of them has 70 nodes with 256 Gb of central memory, which with the associated links has more than 520 Tb. These capacities can be further expanded via direct access to other similar computers available throughout the country. The communication with these computing tools is extremely simple and is done directly via the Internet from a desktop computer.

At the same time, data storage capacities increased exponentially during this period. The only storage available in 1970 was on magnetic tape and was therefore slow to access and limited in volume. Hard disk storage began to develop in 1956, but remained in marginal use for another decade or two. A single disk at that time had a storage capacity of 5 Mb and that capacity has since increased by a factor of more than a million. My department alone, the interdisciplinary Applied Genoproteomics Cluster, or GIGA, today offers us a secure hard disk storage capacity of 1.5 petabytes (Pb) or 1,500 terabytes (Tb) or 1.5 million Gb or 1.5 billion Mb.

There is therefore almost no limit to the computing power available. Enormous data storage is also present and these capacities in biology are widely used by molecular biology which now makes it possible to sequence billions of base pairs in DNA or RNA. Surprisingly, this acceleration of sequencing capacities tends to be even faster than the storage capacities of hard disks, and serious thought is now beginning to be given to the possibility of using DNA itself as a medium for the long- or medium-term storage of information[5]. This would make it possible to store huge amounts of information in the long-term in a minimal volume. The shortcoming being that the information is not quickly accessible.

1.7. Appearance and development of the Internet

This impressive development of personal computers was accompanied by the emergence, development and finally the systematization of their interconnection via the Internet[6]. The idea of interconnecting computers emerged during the 1960s and involved international collaborations including the United States, Great Britain and France. In 1969, the U.S.

5 https://en.wikipedia.org/wiki/DNA_digital_data_storage.

6 www.en.wikipedia.org/wiki/History_of_the_Internet.

Department of Defense launched the ARPA (Advanced Research Project Agency) project, which led to the development of the ARPANET network, which allowed the transmission of information via a distributed network based on blocks of data traveling independently along different routes; the initial message was then reconstructed on arrival. This counter-intuitive transfer mode a priori was chosen in order to protect the communication from any attempt at intrusion, espionage or destruction.

In the early 1980s, the National Science Foundation (NSF) funded the creation of a network of computing centers connecting several major universities, the NSFNET, which for the first time allowed the instantaneous (or quasi) exchange of information between academic centers. Then, in 1981, a computer network specifically linked universities point to point and was therefore different from today's Internet. Electronic messages (e-mails), possibly associated with small files, were transmitted in a delayed manner after having been entirely encoded in a single block from the starting server to their final destination. This network was very successful (more successful than ARPANET at the beginning) and at its peak had about 500 organizations and 3,000 nodes. The University of Liege where I work and Rockefeller University where my collaborator Gregory Ball was working at the time were part of this network, and we were able to establish electronic contact through this means as early as 1986. This replaced in an extremely advantageous way our exchanges of letters by air (now re-christened "snail mail") by which the answer to a precise question or a modification in a text to be published could not be expected before the two-week mark.

However, the system was still far from having the performance expected today from a basic Internet connection. It was based on a low-powered personal computer communicating through a classic phone line via a MODEM (modulator/demodulator) which turned out to be very slow. The transfer in the network was done at the ridiculously low speed compared to the current standards of 9,600 bauds (units of information transmitted per second). The transfer of 15 pages of text without any formatting took 20 or 30 minutes to pass from one side of the Atlantic to the other, but this was phenomenal progress that would completely change the lives of all researchers.

This system, called BITNET, had its limitations, however, which led to its disappearance in the early 1990s. In parallel, commercial services providing information via networks had appeared in the United States and Australia as early as 1989 and quickly became the main actors of these online information transfers. ARPANET and NSFNET were then discontinued in 1990 and 1995 respectively. The merging of the multiple networks that had been created independently gave rise to the first version of the Internet as we know it today. This fusion of networks was realized around the TCP/IP (Transmission Control Protocol/Internet Protocol), allowing in a standardized way the transfer of information packets between computers. This does not mean that the system has not evolved since the 1990s, but its general principle has remained the same.

Between 1989 and 2004, we first witnessed the development of the first global Internet network, now known as Web 1.0. This was associated with the appearance of standardized information search tools or web browsers, first Netscape and then all the competitors that have developed over the years: Explorer, Firefox, Google Chrome, Safari to name only the best known and widely used.

This was the origin of the World Wide Web (WWW or W3) or simply "the Web" which enables consulting, via a browser, pages of information accessible on specific storage sites. It is one of the most important applications of the Internet, alongside other applications such as e-mail (the successor to BITNET), instant messaging, social networks, file sharing between users and, more recently, telephony or image broadcasting (television, films), etc.

Between 2004 and 2009, the 2.0 version (Web 2.0) of this network developed, characterized in particular by faster access speeds, wider distribution following the generalization of portable computers and other means of access (tablets, smartphones, etc.), greater information storage capacities and a higher speed of transfer. In Web 2.0, much of the available information was generated by users themselves, who made it available to the world.

Over the past three decades, the Internet has revolutionized various aspects of our society including culture, commerce and technology. The result has been an exponential increase in near-instantaneous e-mail communication. Various forms of instant messaging, voice over Internet protocols (VoIP), chat rooms, blogs, video conferencing services and e-commerce sites have also emerged. This growth has been truly extraordinary: the Internet represented only 1% of the two-way telecommunication flow in 1993, but it was already 51% in 2000 and reached a near exclusivity of 97% in 2007.

2

The Profound Modification of Research Conditions

In parallel with the technological developments described in Chapter 1, research tools in laboratories have become more sophisticated and miniaturized, thus facilitating any form of analysis or dosage. In 1970, pH meters, centrifuges, scintillation counters and high-performance spectrophotometers were already available, but assays were still performed in large glass tubes, most of which were non-disposable and therefore had to be washed using laborious dishwashing procedures; multi-well plates had not yet been invented. Manual "automatic" pipettes were a rarity (Figure 2.1), and volumes were often measured with large mouth-sucking glass pipettes or graduated cylinders. These limitations, and others that would be tedious to list here, complicated and thus slowed down the work that was later greatly facilitated by the miniaturization and automation of procedures.

These technical advances have greatly facilitated the work of researchers, and the amount of work that can be done by one person in a given time has increased at least tenfold. Moreover, these developments have made it possible to undertake studies that were previously impossible, particularly in the biomedical field. Unfortunately, they have been accompanied by a profound change in the conditions under which this research is carried out.

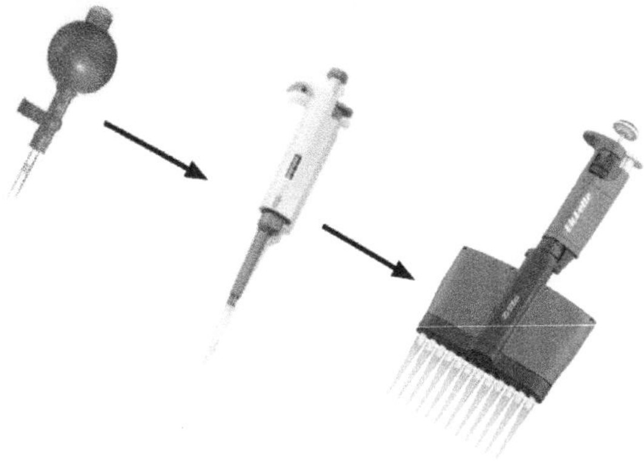

Figure 2.1. *Evolution of laboratory instruments, from the vacuum bulb to connect to large glass pipettes, to the hand pipette and the multi-channel pipette. For a color version of this figure, see www.iste.co.uk/balthazart/fiftyyears.zip*

2.1. The evolution of research funding

In particular, we have witnessed the disappearance or at least the rarefaction of the more than generous funding from which each researcher benefited. Research funding was, at the beginning of the period under discussion, a very marginal concern for a young researcher and involved little or no effort on his/her part.

As an example, my thesis director was at the time widely funded by the university and by the Belgian National Fund for Scientific Research (*Fonds national de la recherche scientifique*, FNRS). He covered all expenses related to my research. This included reagents for radioimmunoassays and other chemicals, and also a huge amount of wheat and corn to feed the 100 or so ducks that I raised annually to study the hormonal basis of their behavior. All these expenses (several hundred thousand Belgian francs at the time) were covered by the so-called "ordinary" credit of the laboratory (money received each year from the university's central administration without having to write an application) and also by a grant from the FNRS attributed to professor Schoffeniels to study the neurochemical mechanisms underlying electrogenesis in the electric eel (*Electrophorus electricus*), a subject of study far from mine. The control of the use of research funds was at the time much less strict than today. This, of course, led to abuses and the financing of my research from this grant was probably one of them.

This abundance of resources also led to a kind of frenzy among laboratory managers in December when they realized that they had too much money left in their regular allocation. Money not spent by December 31 was often lost, and any leftover money would mean a decrease in the funds for the following year. We were therefore buying material that "potentially" could be used in the future, but without being totally sure of it. The system therefore had its shortcomings and implied a certain amount of waste, but at the same time, it allowed a great deal of freedom in research for those who wanted it. I benefited a lot from this situation and in retrospect it was, I think, a good investment. This system of funding research has changed profoundly over the last 30–40 years.

We have progressively moved from a hierarchical system in which a laboratory manager (the professor, head of department) provided all their researchers with the necessary financial means, to a situation in which each researcher has to write grant applications to finance his/her own work. This has inevitably resulted in a major increase in the number of applications for funding per unit of time. At the same time, the amount of funding available dropped considerably, especially if we take into account the significant monetary inflation that was taking place at the time. To make matters worse, the cost of research increased during this period in a significant way as a result of the development of sophisticated but expensive instruments: the new techniques available were increasingly expensive, think in particular of the techniques of molecular biology techniques.

During this period, there was also a growing tendency to replace fundamental research based solely on curiosity (knowledge for knowledge's sake) with research motivated by potential applications in the more or less long-term (knowledge for use). This is funding described as "top-down", where research topics are decided by the funding agencies, as opposed to "bottom-up", where researchers are subsidized to carry out experiments for which they themselves have decided the purpose.

This vision has been adopted by a number of funding agencies. Most of the grants distributed by the European Union have been on the basis of thematic calls asking researchers to solve or at least study a specific problem. However, the creation of the European Research Council (ERC) in 2007 has tended to correct this attitude by creating a fund intended to finance fundamental research. The amount of money available to the ERC is, however, only a small part of the total distributed by the European Union to support research (17% in 2020)[1].

This tendency to control research subjects has, by contrast, been less prevalent in the federal funding agencies in the USA, the National Institutes of Health (NIH) and the National Science Foundation (NSF)), where a large proportion of resources have been and remain largely distributed on the basis of investigator-initiated projects. However, there has been a trend at the NIH over the last decade to request more and more "translational" research, i.e. research directly applicable to the treatment of a disease. This change corresponds in fact to a profound change in society in general, which has gradually lost interest in fundamental research and no longer wishes to finance research with taxpayers' money unless it brings a palpable benefit in the short-term. We will return to this later.

This redistribution of resources has as a result reduced research creativity, and therefore the possibility of making truly fundamental discoveries that will only find an application in an unknown but often distant future. It has also led to the gradual attrition or even the almost complete disappearance of certain fields of fundamental research. For example, in Great Britain, in the 1970s, there were many groups interested in avian endocrinology that were sufficiently funded to generate a large scientific output of international quality. Almost all of these groups have disappeared. The same phenomenon took place in the Netherlands for groups studying the hormonal basis of sexual behavior without necessarily taking a direct interest in the human clinical implications of such studies. All these groups have similarly disappeared. These disappearances could simply be the result of the absence of researchers interested in these topics, but the lack of grants and thus career opportunities in these fields has undoubtedly played a major role.

Finally, it should be noted that the scarcity of research grants has also resulted in a significant decrease in the number of technicians in the laboratories, technicians who provided valuable assistance to researchers. Until the 1970s, the universities were financed by quasi-automatic grants provided by the government, which made it possible to hire research technicians on fixed-term contracts. The change in the financing of universities has reduced these resources and led to the non-replacement and therefore to the progressive disappearance of this support staff. The

1 https://en.wikipedia.org/wiki/European_Research_Council.

researcher who wishes to have a technician to help in the functioning of the laboratory must now find a grant that will finance this salary, and this can only be for a limited period of four or five years at most.

2.2. The current funding situation: multiple sources of grants that are difficult to control

Research funding in Europe and particularly in Belgium is currently difficult to understand. Each country has its own funding agency for basic research (in Belgium, FNRS or *Fonds national de la recherche scientifique*, in France, CNRS or *Centre national de la recherche scientifique*, etc.) in addition to a specific funding organization for medical research (in Belgium, FRSM or *Fonds de la recherche scientifique médicale*, in France, INSERM or *Institut national de la santé et de la recherche médicale*, etc.) and another for applied research in agriculture (in Belgium, FRIA or *Fonds pour la formation à la recherche dans l'industrie et dans l'agriculture*, in France INRAE or *Institut national de recherche pour l'agriculture, l'alimentation et l'environnement*, etc.). However, the distinction between the fields of research that can potentially be funded by these different organizations is not always clear-cut, and the same research can potentially be submitted for funding to several organizations.

Belgium is divided into three language regimes and three administrative regions, making the situation even more complicated. The research funding agencies have been officially regionalized over the last 20 years. The FNRS has a French-speaking branch (FRS-FNRS, *Fonds de la recherche scientifique*) and a Dutch-speaking branch (FWO, *Fonds voor Wetenschappelijk Onderzoek*), and the same applies to applied research (FRIA and IWONL, *Instituut voor Aanmoediging van het Wetenschappelijk Onderzoek in Nijverheid en Landbouw*). These organizations are now completely independent and manage research with different rules using different scientific committees. The regions and communities in Belgium also distribute a lot of research funding according to their own criteria. Moreover, the universities themselves or their faculties distribute grants to their researchers on the basis of local competitions.

In addition, all European countries contribute in varying degrees to a European research funding mechanism, and they have transferred a more or less significant amount of their research funds to it. Researchers from different countries can submit applications to these European funds where they compete with applications from all EU countries. There is thus a very strong competition for these prestigious grants.

This division of the distribution of research funds through many channels has the advantage of allowing most researchers to eventually obtain at least a small grant and continue inexpensive research at a slower pace. However, this system has many disadvantages, including:

– Most of the funds allocated only concern a small amount of money (a few tens of thousands or even a few thousand euros) allocated for a short period of time (one to two years), and there are few mechanisms that allocate larger funds that include operating costs, salaries, travel, etc., i.e. that can support the entire research. This makes it difficult to plan for the long-term.

– To run a laboratory, it is therefore necessary to accumulate multiple grants from different sources that each have their own specific application procedures and different application deadlines spread throughout the year.

– Each researcher is therefore constantly writing a research program or an activity report. This work consumes a substantial part of their time.

– As a result, a given project is often, if not always, financed by several grants simultaneously, and it is impossible to know what has really been done with a given subvention.

– More critically, the peer review of these very numerous applications, in addition to being time-consuming, cannot be carried out in an optimal way, especially in a small country like Belgium. As each organization distributes only a limited amount of money, it is impossible to have specialized scientific panels to review applications in all fields of science. Each panel therefore covers very broad fields and is not very specialized.

– Finally, given the small size of the countries, especially in Belgium, all researchers in a given field end up knowing each other personally or have even collaborated on certain projects, which inevitably creates many conflicts of interest that are more or less well (or not) managed by the various funding agencies. The distribution of certain grants sometimes ends up being based on cronyism or even common political affiliations.

A potential solution to all these problems would be to manage all research credits centrally at the European level. This would involve a large number of researchers and applications that could be evaluated by a large number of competent committees while minimizing conflicts of interest, as is done in the United States for the National Science Foundation (NSF) and the National Institutes of Health (NIH). However, this would not be without its own problems, since the European communities have until now, unlike the NSF and the NIH, funded projects in a top-down manner, reserving only a small part of the funds for projects initiated by the researchers themselves. In addition, the topics covered by the calls for proposals are generally modified every five years, which makes it impossible to obtain successive grants to finance a long-term project. In addition, the European communities have often distributed the bulk of the funds in a preferential or exclusive way to international groups that were initially supposed to have three and then five partners from different countries, but often ask for consortia made up of even more laboratories located in several different countries. All of this makes the organization of grant applications very cumbersome while at the same time greatly limiting creativity.

2.3. Ever-increasing constraints

In addition to having to deal with the financing of their research group, which implies writing numerous grant applications, researchers have had to deal with various and increasingly constraining management tasks over the last three or four decades. These constraints can be grouped into three main types with different origins and purposes: the administration of research and its grants, risk management and the protection of animal welfare where research uses animal models.

2.3.1. *Administrative tasks*

Our society has in recent decades developed an almost morbid need to control all aspects of its operation.

We have thus seen the growth within universities and research funding organizations of a body of administrative staff responsible for keeping up to date various directories reporting on the use of research grants. It is, of course, normal for society to demand accountability for the use of taxpayers' money, but this control has now assumed worrying proportions. Any grant obtained after a complex and often multiple application process will lead to a request for a description of this grant in one or more databases. A proposal for the use of the budget will have to be introduced in the university's financial management system and updated every year. An activity report will be requested each year and at the end of the contract. These reports should, of course, describe the progress of the research and a list of the personnel who worked on the project, whether or not they were paid by the subsidy in question. Some agencies also require a breakdown of the time spent by each staff member on the project on an annual basis, which is reasonable, and also in the most extreme cases month by month, or even week by week or hour by hour, which implies a complex management of the activities[2].

Each publication produced with these grants will have to be entered into various databases (managed by the funding agency, by the university) and today in various archiving and promotion systems on the Internet where it is now necessary to be present such as ResearchGate[3], Academia[4] or Google Scholar[5]. All of these tasks are of course time-consuming and reduce the amount of time left for research.

It has even come to the point of making it compulsory to enter every scientific mission project in a specialized and rather complex database. This is a prerequisite for being reimbursed for expenses, even though it is based on a grant that has been obtained by the researcher. It is supposedly necessary for the university to have complete information on all the missions carried out by its members… Why? In order to generate other reports on activities whose usefulness seems quite problematic, except to allow the university to shine in ranking competitions that are as inept as the rankings of researchers based on their cumulative impact factor (see Chapter 5). Administration seems to have become a self-sustaining and self-justifying activity in the university as elsewhere in society. The University of Liège now employs hundreds of people dedicated solely to administration, probably more than 500 if we combine the central administration and the faculties, for a total of 5,666 staff members and about 12,000 direct jobs. This raises questions.

2.3.2. *Risk management*

Biomedical research today uses many reagents that are potentially dangerous either to the environment or more directly to the researcher

2 www.amplifund.com/blog/how-to-comply-with-time-and-effort-documentation.

3 www.researchgate.net/.

4 www.academia.edu/.

5 www.scholar.google.com/.

handling them. This was already somewhat the case in the 1970s, but the problem has greatly increased since then. We now distinguish between risks related to the use of radioactivity from chemical and biological risks.

Radioactive markers have long been used in laboratories, notably to perform assays or identify metabolic pathways. This use has in fact been progressively reduced thanks to the appearance of alternative techniques using other markers, enzymatic or fluorescent. However, the use of radioactivity remains essential for certain studies. Safety controls at this level were not very present in the 1970s and even though very few (or no?) health problems resulted for the researchers themselves, it must be recognized that the way in which radioactive waste was disposed of led to a certain amount of pollution of the environment by sources with low radioactivity or short half-life radioactivity. All this is now well controlled by a specific body that must legally be present in each university, but depending on the institution, certain control services can be very fussy and overzealous. This leads some researchers to abandon a specific research requiring a radioactive tracer.

The number of chemicals used in a laboratory has also grown almost exponentially, and some of them are more or less toxic. This use is also highly regulated implying in a non-exclusive way the need of:

– filling out lengthy authorization forms before being able to purchase a product;

– keeping a register of all the products present in a laboratory and keeping under lock and key the most dangerous of them;

– keeping a constant inventory of their use;

– eliminating them and eliminating any samples that may have been contaminated by them in a very specific way.

Again, these measures are often common sense and are controlled in all universities. However, they imply a "loss" of time that is sometimes not negligible.

Finally, biological research now uses viruses in a fairly regular way. They are used to specifically modify a certain type of cell, particularly in the brain. It is in this way possible to suppress the expression of a given protein (the receptor for a neurotransmitter, for example), or induce the expression of such a receptor that was not present there, or force these neurons to express a light-sensitive molecule that will activate or inhibit the electrical activity of these neurons when exposed to light of a given wavelength transmitted by an optical fiber (see section 1.5). These viruses are of course genetically modified in such a way that they fulfill the desired function, and it is also ensured that, in principle, they are no longer dangerous for humans. It is essential that this last condition is strictly controlled. This use of viruses is therefore under the control of a specific security service which, depending on the type of virus, will impose more or less restrictive working conditions. The same applies to researchers who study in the laboratory the pathogenicity of viruses or bacteria responsible for diseases. These manipulations must be done in secure laboratories with an increasing level of safety (L1, L2, L3) as more and more dangerous microorganisms are handled. The most dangerous viruses, such as the SARS virus, which is responsible for Covid-19, are handled exclusively in L3 laboratories, which are under permanent depression to avoid any leakage. They can only be entered through double doors, the second of which can only be opened after the first has been closed. Researchers must also wear a full suit in which breathing is done through a filtered air source. The exit requires a shower and a complete change of clothes. Access to these L3 laboratories is only possible after a specific training explaining all these safety measures and their meaning.

The services controlling chemical and biological safety are generally quite sensible. They establish a reasonable assessment of the risks and impose proportionate safety measures while trying not to penalize research too much. However, these controls are sometimes disproportionately zealous in relation to the risks involved. I will simply cite as an example the current attitude towards paraformaldehyde or formalin which is widely used as a histological fixative. In my childhood, I could buy formalin freely in the pharmacy and use it to preserve various small animals that I collected from the wild. My bed room was cluttered with jars filled with formaldehyde containing fish, snakes or other spiders. A slight smell of formaldehyde was constantly present in my bedroom. I may have been lucky, but I am still doing very well for an "old man" of my age. Currently, the product is only handled under a chemical hood, often with a mask and gloves, as if it were extremely toxic. Limiting exposure is probably beneficial, but the current measures seem to me to be a bit excessive.

2.3.3. *Animal welfare*

Our attitude towards animals has changed profoundly over the last 50 years. For a long time, animals were considered as machines with sophisticated reflexes but without consciousness or thought. This view was popularized in France by the writings of René Descartes in the 18th century[6]. Under the influence of the research of ethologists who studied animal behavior, often in its natural environment, this vision changed profoundly during the 20th century. It is now known that animals, at least some of them such as monkeys and corvids, are able to solve complicated tasks, to use and even to make tools and to plan their behavior according to foreseeable future events. Research by primatologist Frans de Waal has also shown that some animals are empathetic (De Waal 2010): they are able to perceive and feel the emotions of other individuals of the same species[7]. This ability was first demonstrated in primates, but it has also been identified in a rodent species, the prairie vole (Burkett et al. 2016).

The very clear line that we thought should be drawn between animals and humans from a cognitive point of view has thus become much more blurred. The more research progresses, the more we realize that at least some animals have much more elaborate cognitive abilities than we thought. There is now a more or less broad consensus that animals are beings capable of feeling pain. This of course raises the ethical issue of how legitimate it is for the human species to exploit animals for food, clothing and, in the case of this book, for scientific experiments.

Various animal protection organizations are calling for a total ban on animal experimentation and petitions have been filed in this regard, particularly in Europe. However, this research remains necessary to make progress in the fundamental knowledge in biology and to validate the efficacy and safety of clinical treatments. The complexity of the functioning of organisms, including the human species, cannot be fully reproduced by in vitro models or computer simulations, called in silico.

Associations opposed to this research, such as the SEA (*Suppression des expériences sur l'animal*), GAIA (*Groupe d'action dans l'intérêt des animaux*) or Antidote Europe, defend the idea that this research on animal

---

6 https://en.wikipedia.org/wiki/Animal_machine.

7 https://en.wikipedia.org/wiki/Frans_de_Waal.

models cannot be transposed to humans. According to one of their statements, the analysis of more than 25,000 articles in life science would indicate a clinical transfer rate of 0.004% (Contopoulos-Ioannidis et al. 2003). The arguments developed by the SEA administrator are nothing new and have already been discussed, dissected and refuted many times[8] (Balthazart et al. 2019). Nevertheless, in view of the importance of the subject, it is important to recall here some fundamental facts, because the therapeutic successes due to animal experimentation, which benefit families faced with the drama of the disease, are real.

This rather basic study, estimating a transfer of data from research to the clinic at 0.004%, actually tests whether an isolated article leads directly to an innovation. However, this assumed frequency in no way demonstrates the inefficiency of fundamental research but the impact of an isolated scientific publication. This statement does not understand the complex process that leads to concrete applications of basic research; it is, at best, a sign of a total lack of understanding of research and at worst of intellectual dishonesty. Hundreds of well-documented examples demonstrate that animal experimentation has been indispensable to the improvement of human and veterinary medicine. For example, no vaccine, human or animal, has ever been developed or validated without animal testing. Similarly, the successful treatment of leukemia is the consequence of fundamental research, a priori unrelated to this subject, on the structure of a family of protein kinases (Druker et al. 2001a, 2001b). And examples abound by the hundreds.

A simple examination of the table of contents of the top international scientific journals (*Nature*, *Science*, *Cell*, *Neuron*, etc.) still shows the predominance of work using experimentation on animals. It should also be remembered that these tests on animals are only carried out following in vitro (cell culture) and in silico (mathematical modeling) tests carried out in the first phases of a given research project. This is a legal obligation. However, the most sophisticated mathematical models are unable to account for the complexity of the interactions underlying the functioning of the nervous system or the immune reactions of an organism, and this situation will probably not change quickly, given the complexity of life.

8 www.lesoir.be/235617/article/2019-07-09/lexperimentation-animale-toujours-une-necessite-pour-la-sante-animale-et-humaine.

No one disputes that animal model testing is not perfect, but it is the best we have to protect patients. The list of the few specific cases in which these tests have not proven to be predictive cannot obscure the vast majority of cases in which the animal study has been able to predict the reactions to be expected in humans following a pharmacological or other treatment. Therefore, experimentation in animals is still a necessity for human and animal health. Veterinary medicine, which every person in charge of a pet is happy to call upon in case of illness, was built up in particular thanks to animal experimentation. And the health of farm animals and consumers is closely dependent on the quality of the veterinary follow-up of animals.

Associations opposed to animal experimentation insist on the existence of alternative methods, but these are nothing more than the in vitro and in silico approaches already mentioned and widely used. They suggest that scientists could do without animal experimentation but that, for obscure reasons, they would refuse. While complementary methods to animal testing do exist, they are currently unable to completely replace animal testing and are unlikely to do so in the short or medium term. As the Brussels Commission on Animal Experimentation (*Commission bruxelloise de l'expérimentation animale*) pointed out in a recent report on animal toxicology testing, "the validation of an alternative method is a long and costly dynamic process that can currently only be applied in specific cases. [...] The validation process can easily take 10 years and 15 years for regulatory approval". This observation was already made in 2010 by the European Commission in response to the European citizens' initiative Stop vivisection, which called for a ban on the use of animals in research. The Commission concluded that "animal testing remains important to protect the health of citizens and animals and to preserve the environment".

Finally, these "animal rights associations" are also protesting against the actions of a lobby of "animal experimenters" who systematically oppose any attempt to regulate or reduce animal experimentation in Belgium. According to them, the university researchers who defend animal experimentation are too closely linked to "big pharmaceutical and chemical companies" (such as Janssen, GlaxoSmithKline and Bayer) which have commercial interests to defend and therefore cannot have an objective opinion on the issue. It should be noted that, by definition, a lobby is a group of people created to promote and defend private interests. However, the scientists who advocate the usefulness of animal experimentation do not benefit personally from it and are not linked to large companies, contrary to what is regularly suggested by these associations. They are researchers paid by research organizations (FNRS, CNRS, INSERM, etc.) or by a university, and their position or salary does not depend in any way on the practice of animal experimentation. If they mobilize to "defend" animal experimentation, it is because they know its methodology, its value and because they consider that it remains indispensable for fundamental or preclinical research. In fact, the majority of researchers agree with the idea that reducing animal experimentation and suffering for moral reasons is laudable.

However, the societal choice of abandoning animal research altogether must not be sold to the citizen by suggesting that research would retain the same quality or even be improved as some organizations fallaciously suggest.

The legislator is well aware of all these arguments and has therefore not yet notified a ban. Very recently (February 13, 2022), a vote in Switzerland showed that, in general, the population is not in favor of a complete ban on animal research[9]. However, legislation has been in place for several decades, both at the level of the European Community and at the level of the member states, which provides a very precise framework for animal testing. These laws are designed to limit the use of animals for scientific purposes and, when necessary, to limit the animal suffering involved. They are based on three fundamental principles known under the name of the 3 Rs for replacement, reduction and refinement (Tannenbaum and Bennett 2015). The laws thus require that animal testing be *replaced* with another form of study (e.g. in vitro cell culture) when possible, that the number of animals enrolled in each experiment be *reduced* to a strict minimum, and that their use be *refined* through various measures such as anesthetics and the use of pain-relieving drugs or through the preferential use of animal models such as fish that have a less developed nervous system.

The application of these measures is strictly and consistently monitored by a series of structural measures and control bodies. Any researcher who wishes to undertake experiments using animals must first follow a legal training course of approximately 90 hours, during which they will learn about the legislation in force in their country. To undertake a given experiment, they must then write a request for authorization according to a

9 www.admin.ch/gov/fr/accueil/documentation/votations/20220213/initiative-populaire-interdiction-de-l-experimentation-animale-et-humaine.html.

precise protocol and fill out a 15-page form describing the planned experiment, the exact number of animals to be used and the manipulations to which they will be subjected. This protocol will be evaluated by an Ethics Commission for animal experimentation which must be legally constituted in each university for Belgium or at the national level for France. This commission will analyze the scientific justification of the research, the justification of the number of animals involved and the details of the procedures envisaged. Modifications may (often) be requested, and it is only after formal approval by the Commission that the researchers can obtain the animals to be used and begin their research. A retrospective report must be drawn up after the experiment has been carried out to evaluate its conformity with the authorized protocol. The researcher will also be required to provide an annual tally of all experimental subjects used. All these reports will be centralized at the European level to establish annual statistics on the use of animals in experiments.

Control bodies have been set up both within the universities and at the national level to verify that all these requirements are met. In my university, this commission on the ethics of animal experimentation was created in response to the law promulgated in Belgium in August 1986, and I have been a member of it since its creation. I even became its president about 15 years ago. It is obvious that all these rules are justified and guarantee, as much as possible, that animal experimentation is carried out under ethically acceptable conditions. However, I would like to emphasize that the respect of these rules imposes on the researcher a quite substantial amount of work which is added to the various constraints described above. It should also be noted that the evaluation of all the ethical protocols and the control of their respect must be ensured by other researchers, often volunteers, which increases the work that has been added over the last decades. This task of evaluating experimental protocols, including their reading and discussion in plenary meetings with about 10 colleagues, takes each member of the ethics committee one to two working days per month!

In summary, we can see that the technical advances that have made research easier have been associated with a number of modifications in the working conditions that have made it more difficult. The demands placed on the researcher have also increased in parallel, as we will see in the following sections.

3

The Computer and its Consequences in Terms of Work

3.1. Changing the flow of scientific information

In the 1970s, scientific research was published in all major European languages: English, German, French, Italian, Spanish, etc. Scientific conferences were also multilingual with or without simultaneous translation, which singularly complicated the exchanges. The English language had not yet become the language of science, which gradually became the rule in the following decades.

This change represented a double-edged sword. On the one hand, it facilitated exchanges between researchers whose mother tongue was not the same by providing them with a communication bridge. But, on the other hand, it required non-native English-speaking researchers to acquire a fairly sophisticated knowledge of the language and to use it in all their written and oral communications. This is a constraint that greatly affects researchers with limited language skills.

In addition, keeping abreast of the progress of research in a particular field was a tedious task occupying a substantial portion of a scientist's time. The Institute for Scientific Information founded by Eugene Garfield (since taken over by Thompson Reuter, and now under the name Clarivate Analytics)[1] published on a weekly basis a 100-page journal printed on bible

1 www.en.wikipedia.org/wiki/Eugene_Garfield.

paper in A5 format that contained the table of contents (a list of all article titles, author names and addresses) of all recently published journals in the field of biology and medicine, the *Current Contents*. Other versions existed for other fields of research (e.g. chemistry and physics). This way, we could get regularly updated information on what had been published in the previous few weeks.

Any university library had to buy this journal, which was consulted in sequence by all members of the laboratory. Reading it took from one to a few hours (depending on the diversity of subjects of interest to the reader) and provided a list of recent articles to be read. Two possibilities were then offered to the researcher to consult an article. Either the library of their laboratory or of another laboratory of their university bought the scientific journal; it was then sufficient to move to this library and to read the document concerned or possibly to make a photocopy of it, if it proved to be particularly interesting. Note that this last possibility only appeared towards the end of the 1970s. Before that, we were simply obliged to take notes. Alternatively, we had to fill out an interlibrary (interuniversity) loan request or write a small card and mail it to the author of the article asking for a copy. Scientific journals at that time printed separate copies of articles, and these copies called "offprints" could be purchased by the author and distributed by mail on demand. The process, as one can well imagine, was relatively slow, and it sometimes took a month or two from the time a new article of interest was identified to the time it was accessed. We conserved these reprints preciously!

If, on the other hand, we wanted to start a research on a new subject and to know the state of knowledge on this question, it was necessary to start a tedious bibliographical work with the help of the *Biological Abstracts*. From the middle of the 1920s, a periodical publication had appeared, containing every two weeks the abstracts (summaries in 200–300 words) of all articles published during the previous period in the field of biology. Each abstract, bearing a specific number, was associated with a series of keywords which were printed in alphabetical order at the end of each volume and associated with this number.

The work then consisted of (a) deciding on a few keywords that would characterize the field to be investigated, (b) collecting at the end of the volume the numbers of all abstracts associated with the first keyword, (c) doing the same for the second, third, fourth, etc. keyword, (d) cross-referencing all these numbers to see which abstracts related simultaneously to all these keywords and finally, (e) consulting these abstracts by leafing through the hundreds of pages in the volume. If the abstract was interesting, we were then in the situation described in the previous section on *Current Contents*: library reading, interlibrary loan request or reprint request.

Each volume of *Biological Abstracts* had 5,000 abstracts on average. More than 250,000 abstracts covering 4,200 scientific journals were added each year (now more than 400,000), creating a collection of more than 11 million documents. To get an idea of the advancement of science over the last four to five years, it was necessary to consult the 24 annual volumes for four to five years, i.e. more than 100 volumes in total. It is easy to understand that this was a long-term task that could take from one to several days depending on the complexity of the research undertaken.

This work is now completely computerized. The *Biological Abstracts* have been digitized and can be consulted online with an automatic keyword search. Many other databases have appeared and can be easily consulted by all researchers. Some are the property of scientific publishers (e.g. Elsevier, Springer, etc.) and their consultation is therefore subject to a fee, but others have been developed by public organizations such as the National Library of Medicine (NLM) and are freely accessible via the Internet to anyone, whether or not they are associated with a university. In the biomedical and biological field, the most comprehensive and widespread database is PubMed[2] which includes all publications in the field since 1966, a total of more than 32 million citations, as well as selective coverage of older publications since 1865.

With a few clicks on a computer, we can almost instantly obtain a list of all the scientific articles that have been published on a subject since 1966 (the beginning of this database), a list associated with the abstracts of the

2 www.pubmed.ncbi.nlm.nih.gov/.

publications and a list of all the articles that are cited in the target article. Links are also present, redirecting the reader to the full text of the articles in the form of an electronic file that will be either open access or in other cases accessible by paying a rather exorbitant fee of a few dozen euros.

The consultation of articles has of course also changed considerably. We have gone from consultation in the library with the occasional note-taking, to making photocopies either in the library or by requesting an interlibrary loan, and finally to consulting an electronic file from any location (laboratory, home, hotel while traveling, etc.). All scientific publishing houses (and many other communication actors) progressively agreed around the year 2000 on a standardized format for electronic archiving, the PDF – for "portable document format" – and started to provide PDF versions of published articles. Most of them also digitized their previous publications so that it is now possible to obtain via the Internet a PDF copy of an article from the 1950s or earlier. A journal such as *Science* published by the American Association for the Advancement of Science (AAAS) is now available in electronic form since December 1889!

This extraordinary acceleration in the distribution of information has also made participation in specialized scientific conferences somewhat less critical. This used to be the only way to find out about the research being carried out at any given time and the results obtained recently. These are now available in the form of publications much more quickly, and the conferences often only present data that has already been published or review summaries that are of course still of interest. However, the "social interaction" aspect remains of great importance, and this has become clear since the Covid-19 epidemic, which has transformed almost all "face-to-face" conferences into video conferences on Zoom or other Internet discussion software. It is indeed by discussing face to face with colleagues, often over a drink or during a friendly dinner, that we learn the little secrets of a laboratory: which experiment did not work? Which aspects are limiting a given technique? What project is currently underway? Who is suspected of manipulating or, worse, fabricating data and whose scientific production should therefore be viewed with suspicion (see Chapter 7)? In these conferences, collaborations and even friendships are created that would otherwise not develop, and all this has a great influence on the progress of science.

Another consequence of this rapid flow of information was to bring some researchers out of their geographical isolation. This was clearly my case. In the 1970s, I was the only researcher working in Belgium in the field of behavioral neuroendocrinology and I also had very few colleagues in Europe, this discipline having developed mainly in the United States. To obtain an answer to a specific question, there were only two possibilities for such a researcher: either write a letter that would not be answered for at least a fortnight or wait for one of the two or three conferences he was attending each year to meet a competent colleague. E-mails and then the Internet have completely changed this situation, and we can now immediately get in touch with any colleague in the world who works in the same field and potentially has the answer to the question. These contacts can be made in writing via e-mail, but for the past decade also via videoconference which really allows for an in-depth exchange.

3.2. Scientific information management in laboratories

The extraordinary development of desktop computers has had a profound effect on the way all of us work. Access to information has been dramatically changed and accelerated. A published article is picked up in less than a few days by scientific databases and becomes accessible all over the world as long as you have a computer connected. More recently, we have even seen the development of directories that include first drafts of articles before they are submitted to the evaluation of colleagues and published in recognized journals. This is notably the case of BioRxiv (read "bio-archives")[3] in the field of biological sciences. These preprints allow access to research results as soon as they are available. It is important to realize that the quality of these results is not guaranteed by the peer-review process and that we should be particularly critical when consulting one of these preprints.

In the past, every conscientious researcher used to keep in their office a mass of photocopies of articles and offprints more or less well classified, which have now been replaced by computer folders containing PDF files or

3 www.biorxiv.org.

even direct online consultation. The organization of information about the literature has been completely transformed.

a)

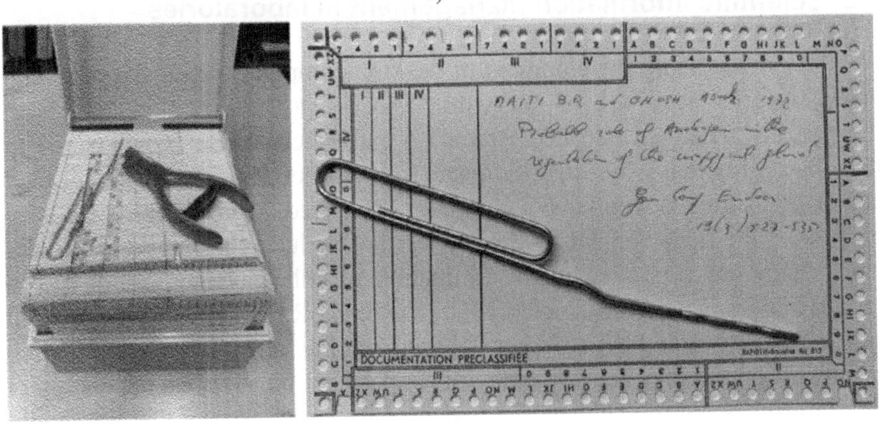

b)

Figure 3.1. *(a) Cardboard sheets and then (b) punch cards used to compile the bibliographic information with the punching tool, the notches and the unfolded paper clip used to select the cards. For a color version of this figure, see www.iste.co.uk/balthazart/fiftyyears.zip*

Initially, light cardboard A7 cards (74 x 105 mm) were used on which references were written or typed, one per card. These cards were then arranged alphabetically by author name or by keyword (Figure 3.1a). A first improvement consisted of “mechanizing” the searches by using cards with bibliographic information but with holes around their edges. According to a personal code to be chosen, each hole could be associated with a keyword, and with the help of a special pair of pliers, the research could cut the outer edge of the hole. Using a thin iron rod, such as a knitting needle or a large unfolded paper clip, we could then thread all cards through the hole concerning the keyword and by shaking make all cards (and references) corresponding to this keyword fall out of the pack (see Figure 3.1b). The sorting process could then be repeated if more specific information was desired associated with multiple keywords. These punch cards made the work a little easier and were used for a few years, but they were soon abandoned when personal computers arrived!

Specific programs for encoding and sorting bibliographic references then rapidly became widespread (e.g. Reference Manager™, EndNote™, Scopus™). They make it possible to encode each reference in a computer file and to associate it with all the information considered useful, i.e. in a non-exhaustive way: the name of all authors, the title of the article, the journal, its volume and the corresponding pages, any potentially associated keyword, and in more recent versions, the abstract of the article, possible personal notes and one or more links to a computer copy of the article stored in the user’s computer or present in online databases.

The programs then allow any given reference to be found by searching any of this information (author, word of the title or abstract, date, journal, etc.), which is obviously infinitely more efficient than any other previous technique. For example, we can remember the name of the senior author (the director of the laboratory) of an article but not the first author who may be a young PhD student.

These programs are now all potentially connected to the Internet, and they enable searching in databases such as PubMed (see previous sections) all the references potentially related to a research topic and downloading in one click all these references in a locally saved file.

These files will of course always be associated with the electronic version of the article, which can be easily consulted. Senior researchers have thousands of photocopies and reprints of articles stored in large cabinets with drawers, all of which are archived on the small cardboard cards described earlier, but, with rare exceptions, they no longer use them. It is now quicker to search for a reference on the Internet and download the corresponding PDF from a database located in the United States than to get up from one's chair, find the card to locate the photocopy and then find it in the corresponding binder!

3.3. Data processing and preparation of scientific publications

This technical evolution has also affected the way in which a scientific article is prepared and published. Traditionally, the researcher wrote a first draft of an article which was then typed by a secretary if the researcher was not able to do it. This text was then submitted to the co-authors, who made changes either by annotations and erasures on the first draft or by using scissors and glue for more important restructuring concerning the general flow of the discourse. A new typing was then necessary, and this cycle could be repeated several times until a version considered final was obtained. Multiple copies then had to be prepared either by photocopying or, in the past, by typing on multiple sheets of very thin paper separated by carbon paper. These multiple copies were then mailed to the publisher of a scientific journal who in turn mailed them to reviewers. These reviewers provided comments and criticisms which were collected by the editor who sent them to the responsible author, everything always being managed by mail. A new version taking into account these comments was then eventually prepared, corrected, re-typed and re-submitted, etc. This cycle could be repeated two or three times and take anywhere from several months to a year.

As an example, the last paper resulting from my postdoctoral stay at Rutgers University (Newark, New Jersey) was written after my return to Liège. Each version of the paper prepared in Liège had to be typed, mailed to Rutgers where my postdoc supervisor, Professor Mei-Fang Cheng, made changes and then mailed it back to me. This took three or four trips back and forth, followed by submission of the paper, evaluation, preparation of a revised version and finally acceptance for publication. This article based on data collected in 1979 was finally published in 1982 (Cheng and Balthazart 1982), a delay that was quite typical at the time.

The computer and e-mail as well as the Internet have greatly facilitated and accelerated this work. The initial text can be written directly on the computer, and all co-authors can make their corrections or suggest modifications directly in the original file that will be sent to them via e-mail. When a final version is obtained, it is now transmitted electronically to the journal in which it is to be published. No need to print multiple copies or run to the post office to drop off your package before closing time. The journal editor can then consult with the reviewers by sending them the text via the Internet and provide a summary of these comments through the same channel. The revision in consultation with the co-authors will be carried out always on electronic files and so on. It is easy to understand that this way of working and communicating saves many weeks and often months before reaching the goal which is the acceptance and publication in a scientific journal.

And the benefit does not stop there, because the accepted article is now already available as an electronic file. The production department of the publishing house can now, within one or two days, produce a version formatted according to the standards of the journal, which will be immediately available on the corresponding website. The availability to the scientific community of research results is infinitely faster today than before the generalization of personal computers and the Internet.

This acceleration also affects research itself. Without even mentioning the various technical improvements that make it possible to collect the results infinitely faster (see in particular in Chapter 1 the evolution of DNA and RNA sequencing techniques), the processing of these results has been completely modified by the omnipresence of computers. Numerous programs for personal computers have been developed and are easily accessible to perform sophisticated statistical analyses of the data. Scientific drawing programs are also available to prepare all types of graphs usually found in a publication, such as histograms, curves, Venn diagrams and pie charts to name a few. These are a wonderful substitute for the earlier technique of drawing in India ink followed by photographing the result and printing multiple copies. Finally, for those using microscopy, the "analog" photography of the samples followed by darkroom printing of multiple copies of the photos to send to an editor has been replaced by digital photography, which is more sensitive, more easily editable and provides a document that can immediately be analyzed quantitatively by sophisticated computer programs and be integrated into the electronic submission of an article.

The work of the researcher has therefore changed considerably. Certain tasks in the laboratory remain indispensable, but all subsequent steps are carried out on the computer. No more long stays in the library, no more laborious calculations with slide rules or pocket calculators of the 1970s, no more drawings with Indian ink, no more analog photography and long sessions in the darkroom. All these works now take place in the same place on the researcher's personal computer. This clearly constitutes a facility and also a certain standardization of the work based very largely now on hours spent in front of a computer screen, which is perceived as constraining for some.

We could imagine that thanks to all these technical improvements linked to the computer, the researcher's work has become infinitely simpler and that therefore the profession of scientific research has become a real pleasure. Unfortunately, this is not always the case, because these technical developments have been accompanied by a very significant decrease of the support staff. The laboratory of general and comparative biochemistry where I worked at the beginning of my career employed two or three secretaries, a draftsman, a photographer and five or six technicians. There was also a mechanical workshop and a woodworking shop that could easily make or repair small research equipment. As a result of severe cuts in research budgets (see Chapter 2), most of these technical support people were not replaced when they retired. As a result, the help that a researcher could have had was therefore lost and they are now supposed to compensate for this loss by using a computer, for better or for worse! The last few decades have also seen a drastic increase in constraints (Chapter 2) and also in expectations, as we will explain in detail in the following pages.

4

The Development of Publishing Giants and Open Access

4.1. The evolution of publishing houses

The last 50 years have seen an extremely significant development in scientific research activity. The number of articles published has increased almost exponentially, from about 400,000 per year in 1980 to 800,000 in 2000 and to more than 1.8 million in 2020, i.e. 5,000 articles per day or 3.5 per minute. Scientific journals have of course adapted and multiplied. Initially, they were often created by scientific societies to promote the development of their discipline. Many new journals then appeared, initiated either by publishing houses (publishers) or individual researchers. As an example, the pharmacologist Matthew J. Wayner initiated in 1966 the publication of *Physiology and Behavior* and then 10 years later of *Brain Research Bulletin*, both of which were later acquired by the Elsevier™ publishing house. From then on, the number of scientific journals grew very significantly, as the publication of original research became a very lucrative business. Thus, we went from 100 scientific journals in 1800 to 10,000 in 1900 and 100,000 in 1990[1]. Since then, the number of creations has been in balance with the number of disappearances.

But at the same time, and somewhat surprisingly, the number of scientific publishing companies has largely decreased, due to a cascade of takeovers and mergers, as observed in all fields of economic activity. A few publishing giants do currently dominate the market, because it is now a market that we

1 https://en.wikipedia.org/wiki/Scientific_journal.

must speak of. In 2001, for example, it was estimated that Elsevier, following multiple mergers, controlled the publication of more than 1,500 journals representing 34% of the main publications in the biomedical field. For the neurosciences, this percentage is probably more than double.

The economic model on which scientific publishing is traditionally based consists of publishing in journals the articles produced by researchers and reviewed by other researchers, all paid by their research institutes or universities; they receive no remuneration for their publications. Authors must transfer all their rights to the publishers and are then unable to manage the distribution and eventual revenues that could be derived from the publications. These journals are then sold to users, mainly university libraries, research institutes and pharmaceutical companies.

Scientific publishing has gradually become an important economic activity, and the presence of giant publishers controlling large parts of the market has allowed them to set prices in a somewhat uncontrolled manner. Thus, at the end of the 20th century, the price of journals skyrocketed, forcing many libraries to abandon some of their subscriptions despite a significant increase in the budget devoted to these purchases. Many actors in the research field have therefore come to consider that the increase in the price of journals had become a serious obstacle to the circulation of scientific information. This is one of the reasons that led to the development of an alternative publication model known as open access. In this model, scientific journals, or very often their electronic version, as there is no longer a paper version, are freely available on the Internet to anyone and the cost of the journals is borne by the authors.

There was another, completely independent reason for the development of open access publishing. Citizens often grouped in associations supporting patients affected by a given pathology asked that the results of research paid for with taxpayer's money be freely accessible, with the implied hope that this would accelerate the implementation of low-cost therapeutic interventions. This is, in my opinion, a rather unfounded motivation, since, on the one hand, articles describing new scientific results are often little or poorly understood by the general public and, on the other hand, they are likely to generate unfounded hopes for treatment. This being said, these requests from patient associations were expressed in an increasingly insistent manner, and this very probably contributed to the emergence of this new publication model.

4.2. Open access: advantages and disadvantages

Numerous journals using this open access model have been created and have rapidly expanded. A substantial amount of the new information produced by scientific research is thus openly available to the research community as well as to patient associations and society in general. This has broken the quasi-monopoly of information held by some publishing giants and forced them "to some extent" to control the incessant increase in their selling prices. This model is not without its drawbacks, however.

On the one hand, the author of a scientific article must now pay the cost of its publication. The publication of an open access journal involves a certain amount of work, including the management of peer review which is still a costly activity in this model, and the layout and management of a rather sophisticated website. All of this must be organized by a dedicated publishing staff, which entails costs that must be covered by the researcher via the grants received. The invoices are non-negligible and amount on average to €1,500 or even €2,000 per article and sometimes up to four or five times more (more than $11,000 for an article in *Nature*). This is only a limited problem for researchers with comfortable resources, but becomes essentially impossible for a less well-off researcher, as is very often the case in emerging countries or in fields benefiting from lower funding than in the biomedical field. They therefore benefit from free access to certain journals but are at the same time unable to publish there.

Another disadvantage of this model is linked to the fact that the cost of the publications was in the classical model assured, adding of course a copious profit for the publishing houses, by the libraries of the research institutes and universities and also by the pharmaceutical industry which has to keep abreast of the progress of research in order to develop new products. In the open access model, this contribution of the pharmaceutical industry is essentially lost, because while the researchers of these firms read a lot and buy many journals, they publish very little, because their discoveries are essentially patented and do not fall into the public domain.

4.3. Unexpected consequence of open access: predatory journals

Very high-quality journals have been created on the open access model and are now recognized as such by the scientific community. However, the open access publication model creates by its very nature a fundamental conflict of interest for publishers. Indeed, publishers enjoy a turnover profit that is directly proportional to the number of articles they publish in their journal. It is therefore in their interest to publish as many articles as possible with little or no concern for their quality as long as this does not harm their recruitment.

For some journals or publishers, the quantity of articles published has thus become a factor that plays a significant role in the management of the journal. The publishing house Frontiers, created in 2007 by two Swiss neuroscience researchers, has published more than 200,000 open access articles between its creation and 2020, including 100,000 between 2018 and 2020, thus marking a fabulous acceleration in pace[2]. They have multiplied the number of journals, each of which has a plethoric editorial board, more than 80,000 members in 2018 spread over more than 130 journals published by the group to date. This gives them a more or less honorary title that they can add to their *resumes*, but in exchange, they have at least a moral obligation to submit or recruit articles for these journals.

This is also the policy of MDPI (Multidisciplinary Digital Publishing Institute), a company founded in 1966 and based in Basel, Switzerland, which publishes over 300 scientific journals to date. These two groups of journals have acquired a certain reputation in the scientific community, but the quality of the peer-review process they organize has been questioned. These journals intensively organize special issues devoted to a theme and recruit guest editors for this purpose, who are then tasked with finding authors to fill the special volume. Articles published in special issues focused on a single theme are known to attract a greater number of citations (see Chapter 5) and thus give a somewhat disproportionate visibility to the journals concerned.

2 www.frontiersin.org.

In addition, many journals have appeared that publish just about anything and everything. These journals have been called predatory journals[3]: they publish scientific articles, almost always on the Internet exclusively, taking the form of an official journal but without presenting the expected guarantees of quality and peer review, in fact they often do not present any. This is literally a scam that takes varying degrees of severity. In the worst cases, authors submit an article that is accepted in an unusually short time, sometimes 24 hours, and they pay the publication fees, but the journal sometimes does not even exist and their text will never be published. In the intermediate cases, the article will be published on the Internet, but the scientific community will not give it any credit due to the nature and reputation of the journal. According to an online site created by an anonymous collective, there are at this stage more than 1,300 predatory journals.

These journals recruit potential authors via extensive advertising spread through more or less personalized and well-constructed e-mails. Many of them suffer from common problems including being addressed generically ("Dear doctor X"), containing numerous spelling or language errors, requesting submission of an article regardless of its nature (original data, literature reviews, comment, letter to the editor) or its length (one or two pages are welcome) within an abnormally short time frame, often a few days, incompatible with the serious preparation of a manuscript, and finally to come from a journal whose speciality often has no relation with the speciality of the targeted researcher. Some requests are, however, a little better constructed and begin by mentioning the title of an article published recently or 10 years ago (!) by the researcher and indicating that its quality has been noticed by the editorial committee (see some revealing examples in Box 4.1). However, these requests never mention publication fees, or they mention exceptional discounts that must be taken advantage of within a very short time. These fees will only become visible after submission of an article.

These messages are considered as phishing or spam and, as such, attempts at scam have multiplied in recent years in an extravagant way to the point of threatening the system of communication between researchers by electronic mail.

3 https://en.wikipedia.org/wiki/Predatory_publishing.

1) Dear Balthazart J,

We are sincerely trying to contact you several times, but there is no response from you, please do not hesitate to respond to this mail.

Kindly let us know your interest, for publishing your manuscript to our Archives of Clinical Gastroenterology (ISSN: 2455-2283).

Submission Date: May 14, 2022.

Review Process Duration: 10 to 15 days.

Releasing Date: End of May 2022.

Please send your paper(s) directly to this mail id.

Await your response.

Best regards,

Michael J, Managing Editor

Problems: computer generated address, English not related to my domain, submission date within 15 days...

2) Dear Doctor. Jacques Balthazart,

Hope our email finds you well.

Appreciating your recent contributions towards the field of Surgery, we are very much honored to invite you to contribute a manuscript for our journal, "Clinics of Surgery (ISSN 2638-1451)" which solely aims to publish and sustain good quality in publishing and to achieve broad international credibility.

The journal is now accepting submissions on a wide range of subjects and disciplines related to the field.

Please submit your manuscript at: clinicsofsurgery.com/submit-manuscript.php

We understand your busy schedule and so please try to submit a short length manuscript such as Case report, Short communication, Clinical Image etc.

We are looking forward to receiving your manuscript soon.

Thank you in advance!

Regards,

Katherine Williams

Editorial Manager

Clinics of Surgery

Problems: address generated by a computer, journal not linked to my domain, accepts any type of article even two pages…

3) Dear Jacques Balthazart,

Hope you are having a great day!

We are in shortfall of articles for successful release of Volume 3, Issue 5. Is it possible for you to support us with your 2-page Opinion or Mini Review for this issue?

We are confident that you will always be there to support us.

Await your optimistic response.

Sincerely,

Ashley Breslin

Modern Approaches in Drug Designing| ISSN: 2576-9170

Problems: journal not related to my field, accepts any type of article even two pages, classic request ("We just need an article to finish a volume")…

Box 4.1. *Some examples of obviously fraudulent article solicitations*

I personally receive several dozen of them per week. They are easily recognized and each one can be removed with a mouse click in a few seconds, but their repetition means that you run the risk of deleting an important message in the process.

Some young researchers may be deceived by these misleading calls to publish either out of naivety or inexperience or consciously, because they are under a great deal of pressure to publish (see Chapter 6) and want to publish at all costs.

4.4. Reactions

The traditional publishing houses have of course reacted to this change in the scientific publishing market. They have transformed some journals with a traditional model into open access journals. In addition, they have introduced for most of their journals the possibility of publishing "article by article" according to the open access model. These are now the so-called hybrid journals. This of course implies the payment of a publication fee often at the same level as what is asked by publishers who are completely open access, around €2,000 or more per article. This is what is now known as "gold open access". It is in fact free for the readers but much less for the authors of articles for whom this model is very expensive.

A part of the academic world would in fact like to see scientific knowledge distributed according to a model known as "green open access" in which the researchers archive their own scientific production in a database maintained by their research institutions or others. Such databases are gradually being set up. The University of Liège created in 2007 an institutional archive called ORBI (Open Repository and Bibliography)[4] in which the researchers of the institution must deposit all their work. This request is in fact mandatory, because only the articles of researchers deposited in the ORBI archive are taken into account in the evaluation of the researcher's activity within the institution.

Following complex negotiations with the publishers, the articles deposited in these archiving systems may be the final published versions, but they cannot then be in open access; however, they may be distributed on a case-by-case basis by the responsible researcher following a specific request sent over the Internet, the electronic equivalent of the reprint requests described above. Alternatively, these may be versions accepted for publication but before the proofreading work (today very limited in most

4 www.orbi.uliege.be/?locale=fr.

scientific journals) and the layout of the publishing houses. Access under these conditions is therefore not entirely free and, for a researcher wishing to carry out a bibliographical study on a given subject, going through the various archival databases of universities represents a rather tedious task.

The generalization of this green archiving would obviously raise the problem of peer review if it is no longer carried out by traditional publishing houses or quality open access journals. This evaluation must be carried out in an organized and rigorous manner, which obviously entails substantial expenses. It would also require a change in the evaluation of researchers, which is currently based largely on the number of articles they publish and, above all, on the prestige of the journals in which their articles are accepted and published (see Chapters 5 and 6).

Centralized commercial archiving systems based on more or less the same principles such as ResearchGate[5], Academia[6] or GoogleScholar[7] have also emerged and are currently an important source of information for the researcher. The updating of all these multiple databases is, however, a significant additional work for the authors. In addition, these archiving systems are in constant conflict with publishing houses who complain about copyright infringement.

Various research funding bodies have also taken a stand on open access and now condition the allocation of their research grants on the condition that all work resulting from this research be archived in an open access database. This is the case of the American NIH (National Institutes of Health) which require the deposit of all publications financed by their credits in the PubMed Central database created in 2000. This database currently includes 7.5 million articles with their full text. This initiative, initially opposed by publishers, was finally accepted on condition that only versions without the final layout produced by the publisher are archived and that these only become freely accessible after a period of embargo ranging from 6 to 12 months. The European Research Council now has similar requirements as the NIH.

5 www.researchgate.net.

6 www.academia.edu/.

7 www.scholar.google.fr/.

In order to accelerate the dissemination of information, systems for archiving manuscripts before their publication (preprints) have also been set up. This system was first developed in the physical sciences but is now also available since 2013 for the biological and biomedical sciences (BioRXiv).

As already discussed, all biomedical publications are of course also included in the PubMed database[8] managed by the US National Institutes of Health (NIH). For each article, the title, an abstract, the name of the authors and their e-mail address are available to request a copy of the article. Links are also provided to the publishers' websites where we can obtain the full text if we have the rights (via a personal or institutional subscription) or pay to obtain a copy, at a somewhat exorbitant price often between €30 and €40.

As we can see, the dissemination of scientific information is actively taking place in complex and constantly changing conditions. Traditional publishers have seen their economic model seriously challenged by open access publishing. They have reacted in various ways to try to maintain their turnover and profits. They have become full partners of the open access version (gold), and they have also set up complex strategies for selling journal subscriptions, including financial agreements, which are obliged to remain secret, with institutions or even entire countries. They now offer subscriptions for non-negotiable groups of publications, which forces some universities to buy a subscription to a journal of little or no interest in order to obtain a subscription to a desired journal.

Complex negotiations have taken place and continue to take place to decide the precise conditions under which articles can be archived in databases (which version, with which embargo, how broadly open is the archive?). A legal war is also continuously underway to try to limit what archiving systems do.

At this point, we may wonder whether universities and their libraries have gained financially with these changes. They still have to buy subscriptions to journals, but in addition, their researchers pay publication fees in hybrid journals (publication in the subscription model or open access) to publish in open access.

---

8 www.pubmed.ncbi.nlm.nih.gov/.

Researchers also find themselves in an uncomfortable situation where they have to carry out literature reviews in various sources (articles available for subscription, open access articles, various archiving systems) and to try to increase the visibility of their work by depositing PDFs of published versions, accepted but unedited versions or preprints in multiple archiving systems.

To conclude this chapter, let us note that since 2011 a website has been developed providing free access to articles collected by various means (often piracy) in databases and from publishers. This site, called Sci-Hub[9], allows us today to consult freely more than 80 million articles. It violates copyright laws in Western Europe and the United States in a rather obvious way, and some publishers, including Elsevier, are in regular litigation with this organization to enforce their rights. Court decisions forced the closure of the original sci-hub.org website set up by Kazakstan-based Armenian researcher Alexandra Elbakyan, but the Internet is an international structure that is difficult to control, so the site keeps rising from the ashes in another country until it is banned again by the courts.

9 https://en.wikipedia.org/wiki/Sci-Hub.

5

The Invention of Journal Impact Factors

5.1. The development of bibliometrics

The progressive digitization of all scientific publications and their instantaneous accessibility through computers has also led to the considerable development of a new scientific activity, bibliometrics: the quantitative analysis of scientific publication activity[1]. This has had major consequences on the functioning of science and the life of researchers.

Until the 1960s, bibliometrics was limited to librarians managing their collections. The Institute for Scientific Information (ISI) was then founded by the American researcher Eugene Garfield and began publishing the *Current Contents* (see Chapter 3), a weekly magazine containing summaries of all the recently published journals in the biomedical field.

Very classically, each scientific publication cites other articles on the same subject and discusses the contribution of the new results in relation to what was already known. In 1955, Eugene Garfield had the idea of using citations in articles to analyze the relationships between them and thus allow researchers to more easily find out who are their published colleagues in a given field (Garfield 1955). Indeed, if an article X cites an article Y, the researcher publishing X is necessarily aware of Y's research, but the author of Y will not necessarily be aware of the existence of X. Similarly, if X and Y are cited by other articles but do not cite each other, then the authors of X or Y will not be able to establish a link and realize that they have a common research topic. The primary goal of this new type of analysis was of course

1 https://en.wikipedia.org/wiki/Bibliometrics.

laudable and facilitated the work of researchers. However, it had negative consequences that are considered below. The first volume of the *Science Citation Index* analyzing this information was published by the ISI in 1963, thus providing a new and very useful tool for researchers. This first publication covered only the natural and medical sciences, but volumes dedicated to the social sciences and the arts and humanities were soon to follow.

These data facilitated access to scientific information and also allowed the calculation of various quantitative indicators describing the evolution of science from a strictly statistical point of view without considering the content or the quality of the publications. One of these indicators is the half-life of an article, which indicates the time after which the citations of an article will have decreased by 50%, a notion similar to that of half-life describing the decay of radioactivity. But in a more critical way, another indicator developed in parallel was to play a major role in the evolution of science and the way it is practiced; this is the impact factor (IF)[2]. This indicator, which was supposed to measure the relative importance of different journals, was initially called the Journal Impact Factor, but the reference to journals was quickly abandoned, which is not without consequence, as we shall see.

The IF for a given journal is calculated each year as the total number of citations over the past two years to articles published by that journal divided by the number of those same articles. It is therefore the average number of citations made to each article in the previous two years. Thus, a journal with an IF of 5 publishes articles that are cited an average of five times in the year of publication and the following year. Among the generic journals, *Science* and *Nature* have very high IFs, ranging between 40 and 50 (47.7 and 49.9 respectively for 2021). Some medical journals have even higher IFs exceeding 50 (e.g. *New England Journal of Medicine*, IF = 91.2 or Lancet, IF = 202.7; values cited for the year 2021, i.e. the latest values available in 2022). It should be noted, however, that most journals have IFs below 1, i.e. their articles are not cited once on average, as shown in Table 5.1.

This quantitative measure of journal usage was originally devised by Eugene Garfield as part of the ISI, which, after several buyouts, is currently in the hands of the Clarivariate company that has been managing the IFs

2 https://en.wikipedia.org/wiki/Impact_factor.

since 2016 and publishing them in the *Journal of Citation Reports*. It has quickly replaced much of the admittedly more subjective but probably better founded assessments of their quality that were based on analysis of the publications themselves, the reputation of the editors and their associates, and the history of the journal. In most people's minds, a journal with a high IF is now considered more prestigious because it is more cited than a journal with a lower IF.

Impact Factor	Number of Journals	Ranking (Top % of Journals)
10+	239	1.9%
9+	290	2.4%
8+	356	2.9%
7+	447	3.6%
6+	610	4.9%
5+	871	7.1%
4+	1,399	11.4%
3+	2,575	21%
2+	4,840	39.4%
1+	8,757	71.2%
0+	12,298	100%

Table 5.1. *Cumulative distribution of impact factors calculated in 2017 according to the Journal Citation Report*[3]

5.2. Disadvantages and limitations

Although the IF calculation is simple and allows for quick comparisons of journals, it has a number of limitations and potential distortions. Some of the main problems include the following:

– The number of citations of an article does not necessarily indicate its quality. These citations can be influenced by various factors such as the interests or the social issues of the moment. For example, the Covid-19

3 Retrieved from www.mdanderson.libanswers.com/faq/26159.

pandemic has obviously increased the consultation and citation of articles on the subject.

– The IF depends to a large extent on the field of research considered, the number of researchers working in it and the number of publications they produce. An article published on cancer will always receive more citations than an article in zoology because there are 10 or a 100 times more publications in the first field.

– Moreover, different fields of science evolve at different speeds. The two-year period used for the calculation is short and therefore potentially suitable for areas of research that evolve rapidly (e.g. biomedical sciences). Other fields evolve more slowly (e.g. human sciences) but will recognize after a few years the essential value of an article that becomes a classic of the field after only 5 or 10 years. Thus, IFs can only be compared within a specific field of research and only to the extent that we can effectively delineate the boundaries of a field. Much research is nowadays interdisciplinary and thus de facto part of several research fields, which makes the value of their IF difficult to estimate.

– A truly innovative discovery will take a long time to catch on and will therefore garner few citations in the two years following its publication. The IF thus tends to devalue true innovation.

– Finally, the IF of a journal can be manipulated by its editors. Various techniques are possible. These include, but are not limited to, the following, in order of importance of the associated ethical problems: a) publishing numerous review articles, which by their very nature are more widely cited; b) for journals that only publish a few volumes each year, grouping their publication at the beginning of the year, which gives them more time to accumulate references; c) encouraging or even obliging the authors of an article during its review to cite articles from the journal in which they are trying to publish. Although these practices are relatively rare, they have been detected and undermine the credibility of this index.

The IF of a journal should therefore be viewed with caution as it reflects only one aspect of its quality. Other quantitative indices of a journal impact have been developed, but they suffer from similar problems. These include the five-year IF calculated on the same principle as the IF but over a period of five years instead of two. Another example is the Immediacy Index, which is quite similar and calculates the number of citations a journal receives divided by the number of articles it publishes in a year. On the contrary, the citation half-life represents the average (in years) of the age of the articles of a journal that are still cited. A half-life of five years for a journal means that articles published from 2016 to 2020 represent half of the citations of that journal in 2020, the other half being before 2016. This index thus measures the persistence of citations in a given field and does better justice to work in slower-moving areas of science.

Various quantitative studies have also shown that the quality of an article, measured, for example, by the statistical power of the study, is not related to the IF of the journal in which it was published. Recall that the power of a study measures the probability that it will detect an experimental effect that does in fact exist. This probability depends of course on the importance of this effect and also on the size of the sample, which should be larger the smaller the effect studied. If this power is low, the probability of missing an effect is high, but more seriously, there is a great risk of reporting an effect that does not exist and therefore cannot be reproduced (Munafo et al. 2017). This is one of the reasons that forces a researcher to withdraw a published article at a later date. There are of course many other reasons for withdrawing a paper, including the manipulation of data discovered a posteriori, and we will discuss this further.

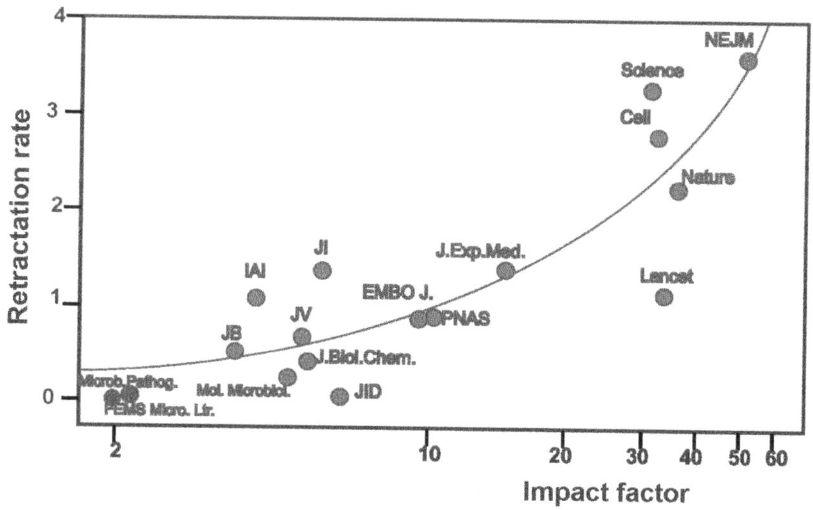

Figure 5.1. *Relationship between the impact factor of a journal and the rate of retraction rate of articles published in it (ordinate). Redrawn from Brembs et al. (2013). For a color version of this figure, see www.iste.co.uk/balthazart/fiftyyears.zip*

In a counter-intuitive but certainly explicable way a priori, it was found that the number of retractions of articles is in fact directly and not inversely proportional to a journal's IF (Figure 5.1). This high incidence of retractions could of course be explained by the fact that journals with a high IF are more widely read and criticized so that a problem with an article will be more easily identified. However, it also raises the question of whether a journal's high IF necessarily indicates a higher quality of the articles published in it.

5.3. Use for the evaluation of researchers

Despite these limitations, IFs and other bibliometric indices have over the past 30 years become increasingly important in the evaluation of research. While the evaluation of the quality of a journal by one of these quantitative indices has its limitations, the problem becomes even more critical if we apply these indices to the evaluation of a given researcher or even to the comparison of researchers among themselves. The research world, particularly in Europe, has been caught up in a frenzy of evaluation and ranking over the last few decades. Bibliometrics and in particular IFs offer an easy, if highly questionable, way of making such so-called quantitative comparisons.

When applying for a nomination, a promotion or a grant, it has thus become relatively common practice to decide on the quality of a researcher on the basis of the number of his/her publications and the IF of the journals in which these articles are published. The productivity and quality of a researcher's work can thus be characterized very simply by the arithmetic sum of the IFs corresponding to each article he/she has published. This makes it possible to carry out evaluations and rankings of researchers in the absence of detailed knowledge of the research field in question. Paradoxically, however, everyone agrees that the evaluation of researchers and their research programs should be carried out by peers who are familiar with the subject and with the quality of scientific journals independently of their IF.

It is also important to emphasize here that the IF reflects an average of the citations of the different articles of a given journal, but the article-by-article figures are extremely heterogeneous and very far from presenting a normal distribution in statistical terms: in general, a very small number of articles accumulate many citations while the others are rarely or not cited.

This aspect has also been analyzed in a quantitative way (Weale et al. 2004). For example, a study of all articles published in the field of immunology (13,125 articles in 105 journals) and surgery (17 articles in 120 journals) during 2001 shows that, on average, only 16.7% of articles (interquartile range (IQR) 13.6–19.2) accumulated half the citations. In contrast, 23.7% (IQR 14.6–42.4) of the articles had never been cited. The distribution of citations in a few journals is shown as an example in Figure 5.2.

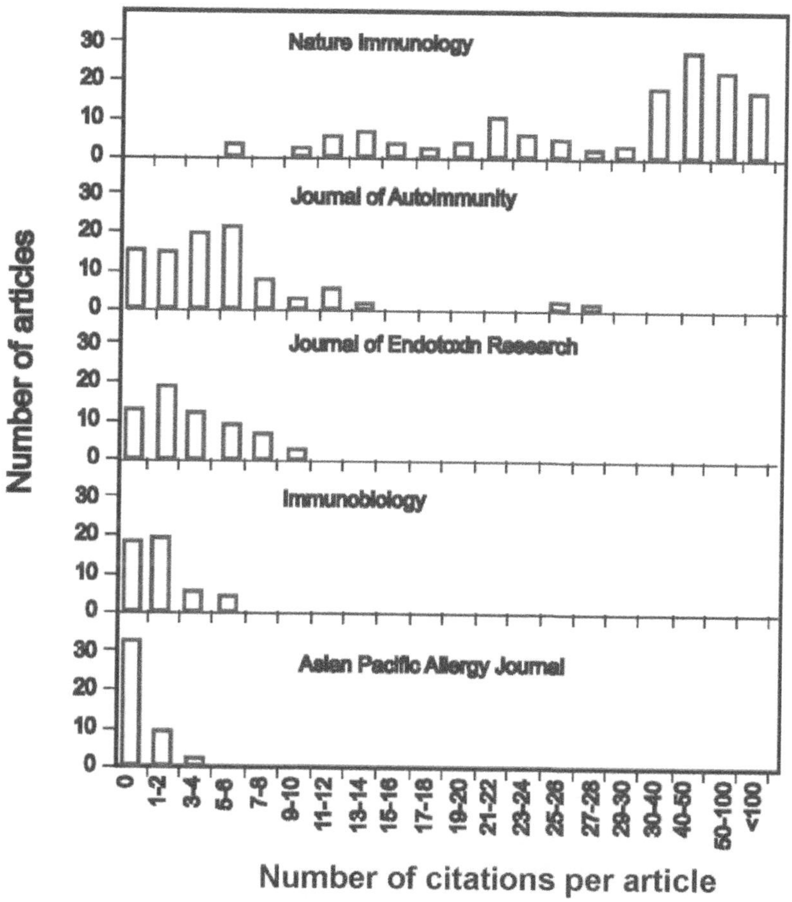

Figure 5.2. *Distribution of citation counts per article in selected immunology journals. Redrawn from Weale et al. (2004). For a color version of this figure, see www.iste.co.uk/balthazart/fiftyyears.zip*

This distribution is extremely different from journal to journal. The distribution of citations in the very high IF journal *Nature Immunology* is very broad, but in the *Asian Pacific Allergy Journal*, which of course has a much lower IF, citations are concentrated on the left-hand side of the distribution with a large number of articles that are never cited. Somewhat reassuringly, according to the same study, the proportion of uncited articles is negatively correlated with the IF in both immunology (Spearman's non-parametric correlation coefficient, rho = - 0.854; see Figure 5.3) and surgery (rho = - 0.924). Thus, high IF journals have a smaller number of uncited articles and the citation distribution of these articles is better distributed. This distribution of citations should therefore be more regularly quantified as it represents another quality variable of a journal that has a direct relationship with the impact of individual articles published in it.

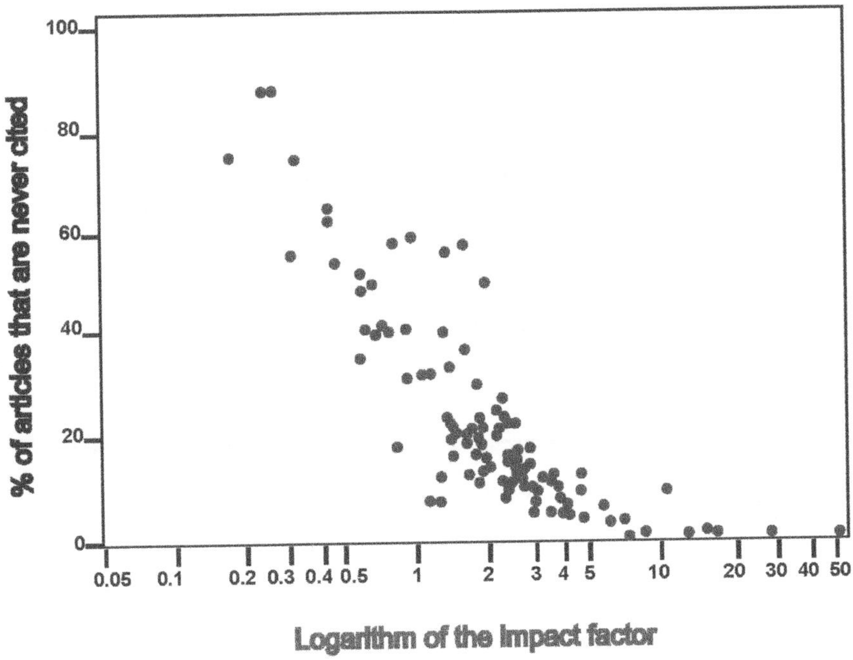

Figure 5.3. *Relationship between the impact factor of a journal and the percentage of articles in that journal that are never cited. Redrawn from Weale et al. (2004). For a color version of this figure, see www.iste.co.uk/balthazart/fiftyyears.zip*

This being said, the IF of a journal does not reflect the number of citations received by a given article published in that journal, and even less its quality.

It is therefore possible to publish in a high IF journal an article that will receive little or no citation. Using this measure to evaluate a researcher is therefore nonsense.

6

The Race to Publish and the Inadequate Methods of Evaluating Researchers

6.1. Increasingly abundant publications

While these quantitative citation indices were being developed, the number of researchers increased dramatically and all of them began to apply for funding, whereas previously only the heads of departments competed. In addition, the amount of money devoted to research reached a ceiling or even decreased in absolute value, so that it was proportionally reduced by a large amount. This decrease was partly linked to the evolution of the general economic situation: we had left the Golden Sixties, during which the economy was growing rapidly. In addition, the cost of research in the biomedical field had increased significantly as a result of the sophistication of techniques. The result was a major decrease in the percentage of grant applications that could be satisfied and a growing difficulty in funding a laboratory.

In the 1970s and 1980s, for example, the various NIH Institutes funded 20–30% of the applications submitted, sometimes as much as 35%. This percentage declined markedly towards the end of the 20th century to stabilize at values around 10 ± 5%. This occurred despite very generous increases in the NIH budget from the US government, even though we compare in constant dollars that are compensated for average inflation (which is less, it must be said, than inflation in research costs). During this period, the NIH budget that was just over $10 billion in 1996 grew to about $50 billion in 2022. This increase took place very gradually except in the years 1998–2003, during which this budget was doubled under the impetus of President George W. Bush. This temporarily increased the percentage of funded applications, but created a huge problem when the increase in the total budget came to a halt while the cost of research continued to increase rapidly. All additional researchers who had committed to an academic career now found it extremely difficult to renew their financial support, creating an enormous frustration and disruption to the system.

All of this has created among researchers a race to publish and then progressively to publish in high IF journals, which, in my opinion, has been detrimental to the quality of science in general. The evaluation mode based on IFs leads to the publication of more and more articles, potentially more or less sloppy, based on few or poorly confirmed observations, and to the "splitting" of important publications into a series of small articles in order to multiply their number. Instead of preparing papers reporting significant advances, the data are organized into the least publishable units (LPUs).

Until the middle of the 20th century, scientific researchers, whether employed full time in a laboratory or working as amateurs, for example in the field of zoology, carried out work and decided to publish the results if they had found something significant. This attitude gradually turned into a race to publish. In a way, publishing became the primary goal of research. Implicitly, we went from a desire to publish to an injunction of the style "publish or perish", then with the generalization of the use of IFs to "publish in high IF journals or perish" and, finally, this attitude becoming generalized and the competition between researchers increasing to "publish often in high IF journals and you may not perish". This evolution has been caricatured (Figure 6.1).

There is of course a positive aspect to this change in attitude. If research has been done, it is important that the scientific community be informed even though the results are not very demonstrative or even negative (no effect of an experimental manipulation). This will prevent other researchers from doing the same unnecessary work again.

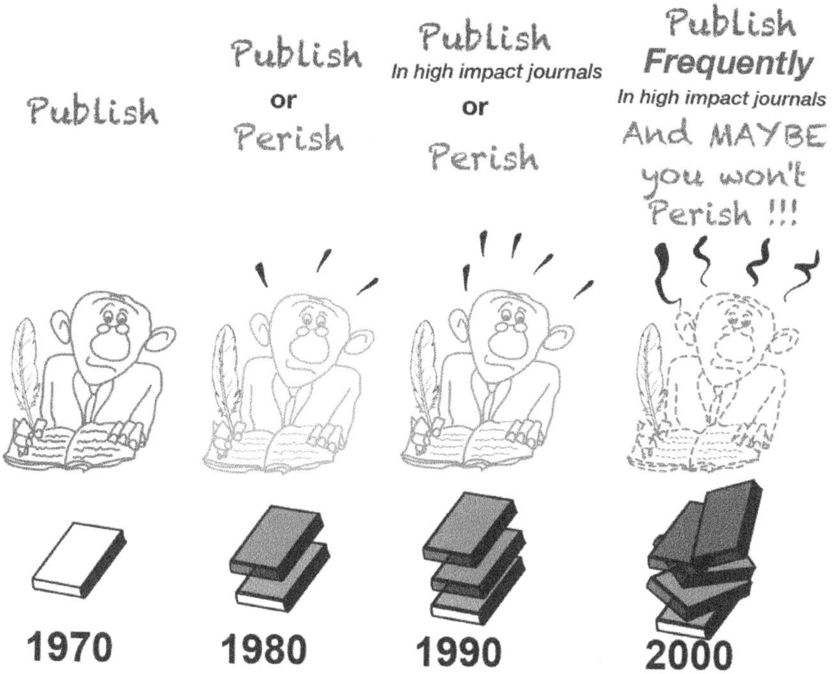

Figure 6.1. *The evolution of the pressure to publish over the last few decades*[1]. *For a color version of this figure, see www.iste.co.uk/balthazart/fiftyyears.zip*

However, this evolution of the publication strategy was linked primarily to the fact that the promotion and grant applications of researchers were increasingly decided, openly or not, based on their cumulative impact factor. More dramatic deviations have even been observed in some countries (Pakistan, China, South Korea, etc.) where financial retributions were directly attributed to researchers who managed to publish an article in a high IF journal. In China, for example, the Beijing Institute of Biophysics awarded a bonus of 2,000 yuan (± €275) per IF point for publication in a journal with an IF between 3 and 5 and 7,000 yuan (± €970) per IF point for an article in a journal with an IF above 10. A paper published in the journal *Science* with an IF of 47.7 in 2021 thus induced a bonus of about €46,250. In Pakistan, researchers can receive a bonus ranging from $1,000 to $20,000 depending on their cumulative impact factor over the year (Fuyuno and

1 Redesigned from an idea by: www.deviantart.com/velica/art/Publish-or-Perish-645355248.

Cyranoski 2006). It is easy to understand why a researcher might be tempted by such a reward and go to great lengths to obtain it even at the cost of violating some rules of professional ethics. Such practices are an indirect incentive to fraud, and engage in plagiarism or selection, even fabrication, of data for publication. We will return to this subject in Chapter 7.

6.2. Evaluation of researchers and grant applications

Numerous researchers and organizations have progressively expressed concern about this drift in values. On the one hand, it has been proposed to use other indices of visibility and productivity of researchers. Thus, we now have IFs calculated over five years (IF_5), which somewhat moderates the discrepancy between research fields that evolve at different speeds. Similarly, we have also started to calculate the citation half-life of articles in a journal, i.e. the time after which the number of citations of articles published in a given journal has decreased by half. All this concerns the evaluation of journals but not researchers. For the latter, the physicist Jorge Hirsch has proposed a supposedly more objective index called the H index that measures the visibility of a given researcher by the number X of articles published by this researcher that have received a number X of citations (Hirsch 2005). For example, a researcher who has published 45 articles of which 20 have been cited 20 times or more will have an H index of 20. This is therefore not an index based solely on quantity (number of publications), nor on a measure of journal impact, but on a composite measure combining the two aspects in a somewhat arbitrary way. Such an index has a number of advantages that have been detailed by Hirsch himself in the cited publication, but whose discussion is beyond the scope of this book. It also has its limitations.

It is immediately apparent that this index can only be used to compare researchers with the same number of years of experience that has allowed them to accumulate publications and their citations. Moreover, these researchers must work in the same field of science, given the disparity in the number of citations from one field to another. This index also suffers from other limitations, in addition to the obvious fact that it relies solely on citation numbers to evaluate the researcher. Is a researcher with five papers cited 100 times or more (H = 5, but total citations > 500) better and more or less visible than a researcher with 20 papers, 10 of which are cited at least ten times (H = 10, total citations = at least 100). Obviously, this index suffers from flaws partly similar to those of the IF. If all these quantitative evaluations based in one way or another on citation numbers allow us to distinguish and probably rank researchers with very different backgrounds (a Nobel Prize winner and a young doctoral student), they are of little interest to compare two researchers at the same stage of their career who are applying for the same position or grant. The problem is even more critical if a university decides to recruit five people, regardless of their discipline. Then we end up comparing a novelist with a physician, a biologist with an engineer, and any quantitative index is meaningless. An evaluation by the peers taking into account the publication activity of researchers and their field of research, in a qualitative and not a brutal quantitative way, is the only way to provide a reliable result.

This obviously raises the question of the amount of work involved in such evaluations following the significant increase in the number of researchers and the number of their publications. One solution that has been proposed is to ask researchers applying for promotion or a grant to list only the five publications that they consider most representative of their work, to ask selection committees to evaluate these publications independently and to base decisions on this review rather than on a quantitative index. This solution is at least partly implemented now in some places.

The significant work involved in writing and evaluating research proposals for funding has become a frequent topic of discussion in the scientific community. Writing a well-argued grant application is time consuming and the same is true for its evaluation. Moreover, due to the intense competition, the percentage of funded applications is often low and is, in many countries and for many funding agencies, often less than 20% (Meirmans et al. 2019). The overall cost is significant. A 2013 study estimated that writing and evaluating applications to distribute a total of 40 million euros cost the sum of 0.5 million euros, which obviously seems unreasonable.

As a result, radical solutions have been proposed. A researcher from Indiana University in Bloomington, Johan Bollen, for example, has proposed that research resources be distributed equally among all active researchers at any given time while requiring them to anonymously redistribute a substantial portion of the funding received to other researchers in their field they consider the most deserving (Bollen et al. 2017). Bollen and colleagues tested their idea through computer simulations and showed that if researchers were required to give back 50% of their allocation to colleagues who cite their publications, research funds would ultimately be distributed in a manner quite similar to what funding agencies currently do. And in this model, researchers would no longer have to write grant applications and funding agencies would no longer have to evaluate them, thus saving a tremendous amount of time and money.

The scientific community was understandably skeptical. Several objections to this model have been presented, including the fact that cronyism could distort the system: a given researcher would transfer his/her portion of the allocation only to friends regardless of the quality of their work or alternatively would systematically avoid funding competitors. One advocate of this system, Martin Scheffer of Wageningen University in the Netherlands, acknowledges these limitations but counters that control algorithms could limit the problem by, among other things, prohibiting the transfer of funds to researchers with whom one has collaborated and published. Scheffer further proposed in 2014 that the Netherlands Institute for Scientific Research (NWO) conduct a real test of this model via a limited experiment. This request was even approved by the Dutch parliament in June 2016 (de Vrieze 2017), but, to my knowledge, was never implemented. However, the NWO is still seriously considering profound changes to the funding system of research and the idea of self-organized fund allocation (SOFA) is still alive[2]. Its benefits and also its potential risks continue to be discussed (Bollen et al. 2019). A limited experiment would likely be worth undertaking.

Most researchers and research funding organizations are therefore well aware of the need to change the way research is evaluated. For example, the Belgian science foundation FNRS has just signed the DORA or San Francisco Declaration on Research Assessment[3], which gathers a set of recommendations on good practices in the evaluation of scientific research. According to this agreement, less weight should be given in the evaluation of researchers to the names (and the IF) of the journals in which their work is published. The FNRS therefore has announced that it will no longer be possible to take into account the IFs of the journals in which the researchers have published, and that they will have to be evaluated on the basis of the

---

2 "Twisted ways of impact – SOFA", KNOWeSCAPE, January 10, 2018 [Online]. Available at: www.knowescape.org/twisted-ways-impact-sofa/.

3 www.sfdora.org.

five most representative publications of their career that they will have provided. It remains to be seen to what extent these good resolutions will be implemented.

Nevertheless, the ideal solution has not yet been found and quantitative indices are still used, despite their shortcomings, especially in Europe, much less in the United States. At this stage, it is clear that this fever of evaluation based on quantitative bibliometrics has had significant effects on the way research is published and on its quality. The damage is done and a full recovery will be difficult.

7

The Consequences: An Overall Deterioration of Research Quality

The pressure to publish more and more in high-impact journals exerted on researchers by the funding system has obviously led them to adapt their publication strategy. This has led, at best, to a deterioration in the quality of the articles and, at worst, to the publication of fraudulent data. The direct consequence has been a decrease in the reproducibility of science, a phenomenon that has been widely commented on by Munafo et al. (2017). The problems present in papers published under this pressure are of two natures that must be distinguished: the publication of rushed or sloppy research often resulting from ignorance or (unintentional) "honest mistakes" by the researchers, on the one hand and, on the other hand, scientific fraud including fabrication or falsification of data and plagiarism which are clearly punishable. These different possibilities will be discussed separately.

7.1. The "bad", not very rigorous, science

The quality and therefore the reproducibility of a scientific study can be dramatically diminished by many factors, often without the knowledge of the researcher, if he/she does not have a solid methodological training. Unfortunately, this training is very rarely incorporated into the curriculum and education of researchers. Researchers have to learn the most basic rules of methodology "on the job" through trial and error. This results in many young researchers conducting experiments in violation of these rules of experimental methodology, which, of course, impacts the quality of the results when it does not render them completely unusable. We will briefly consider in sequence the major methodological problems that affect the quality of science.

7.1.1. *Cognitive biases*

We all tend to see information in the world around us that is not there. Thus, everyone immediately interprets an emoticon based on two small circles, a vertical bar and a horizontal bar inside a larger circle as a human face (Figure 7.1a)[1].

This tendency is the basis of many optical illusions. In Figure 7.1(b), we see alternately a saxophone player or a human face seen from the front depending on whether we focus on the left or right side of the drawing.

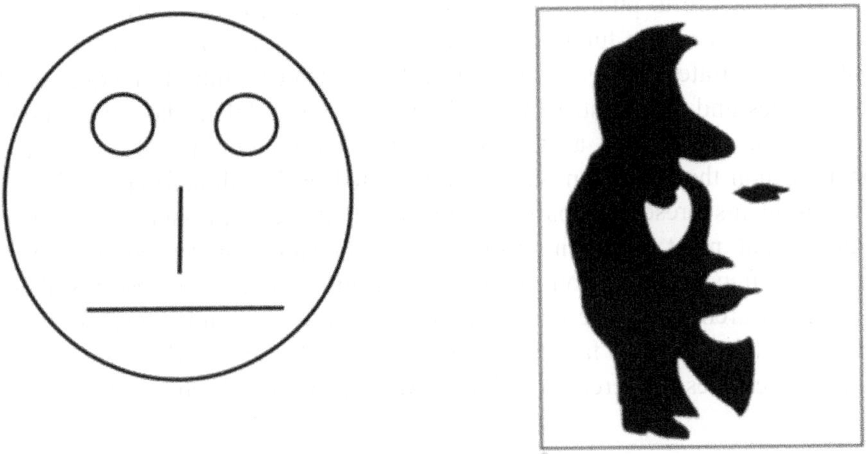

a) b)

Figure 7.1. *Example of cognitive bias*

COMMENT ON FIGURE 7.1.– *(a) Everyone recognizes in the left figure a human face, although it is only roughly suggested. In (b), we will see alternatively a saxophone player or the shadow of a human face according to the point on which we focus.*

---

1 www.en.wikipedia.org/wiki/Apophenia.

Gamblers who play games of chance will, on the basis of this phenomenon, tend to detect logical sequences in datasets (playing cards, lottery results, roulette numbers) where there is no real pattern. In science, the researcher may similarly tend to identify a non-existent pattern in a dataset. In a qualitative analysis, for example, the researcher may tend to believe that there is an experimental effect by focusing on the few data that show it while ignoring those that contradict this idea.

The recipes for avoiding this pitfall are obvious in science. The analysis of results should be carried out in a manner as objective as possible. For example, the researcher quantifying behaviors comparatively in a group of animals subjected to an experimental treatment and in their controls should not know (he/she should be "blind" to) who received the active treatment or the control treatment. Similarly, if we evaluate the number of positive cells for a marker in brain sections, the microscopic slides should be encoded in such a way that the experimenter does not know at the time of the analysis which section belongs to which experimental group. Only when all data have been collected should the researcher discover which experimental group they belong to so that the results can be analyzed. The conclusions to be drawn from these results must then be developed using proven statistical methods (see sections 7.1.5 and 11.3).

7.1.2. *The absence of randomization*

If we want to test the potential role of a given experimental factor (a drug, medication, surgical procedure, etc.) on a given dependent variable (morphological, physiological or behavioral), it is absolutely critical that this factor be the only one that consistently distinguishes the experimental group from the control group.

In practice, however, it will be impossible to achieve perfect equality of all conditions in the two groups. For example, it is impossible to test all the subjects in an experiment at the same time of day: spontaneous daily variations in the variable being tested will therefore contaminate the results. If all animals in the experimental group are, for example, tested before those in the control group, a difference between the two groups could be observed that will then be erroneously interpreted as showing an effect of the experimental treatment when in fact it is only the consequence of daily variations that were not adequately controlled.

The imperative solution that must be put in place is to randomly distribute all the conditions that are not absolutely identical in the two groups so that, on average, they are not different except for the experimental factor being tested. This randomization must, most importantly, concern the distribution of experimental subjects in the two groups. Catching subjects haphazardly in their housing cage is not an acceptable procedure because various studies have shown that the same subjects are always captured first and the others last. Their escape behavior and agility are therefore different, and these differences could be correlated with other differences in the dependent variable that will be studied.

After allocation of the subjects to the different experimental groups, we must then check that the variables that potentially influence the final result (weight of the subjects, age, sex if relevant, etc.) are not different in the two groups. All procedures such as injections, various manipulations, blood samples, biopsies at the end of the experiment will also be done in a random order to avoid introducing any systematic difference between the two groups; for example, a difference linked to the time of day when the manipulation was carried out. Similarly, housing of the different animals in the experimental groups must be organized in a random fashion. It has indeed been shown that the simple fact that mice are placed at different heights in the shelves of an animal house (differential exposure to light, different temperature, etc.) is enough to induce reproducible differences in their behavior. If this were the case, the interpretation of the result of the experiments would be equivocal since a difference between groups could be due either to the experimental factor tested or to a factor that was poorly or not randomized.

This randomization of all conditions that cannot be identical in the experimental groups seems easy enough in a simple experiment with only two groups of subjects or data. However, the situation can become much more complex for factorial designs that test several experimental factors at once. If, for example, we want to test in males and females (factor 1 at two levels) the effect of a drug at three different doses compared to a control treatment (factor 2 at four levels) and compare absorption by the oral route or by intra muscle injections (factor 3 at two levels), the design will include 16 experimental groups (2 x 4 x 2) that must all be matched for experimental conditions (time of drug delivery, order of collection of experimental measurements, organ harvesting, etc.) in order to avoid drawing erroneous conclusions about the differences between groups. This requires careful planning to avoid concluding that one factor has an effect, based on systematic differences in study conditions between groups. This requires careful planning of the experimental protocol before its implementation begins.

If a researcher is unaware of the presence of a poorly randomized factor, he/she will publish results that erroneously conclude that there is an experimental effect. It is clear that under these conditions, these results will later turn out to be non-reproducible, leading to a waste of research resources.

7.1.3. *The use of poorly validated techniques*

The pressure to publish more and faster has occasionally led some researchers to use techniques without assessing various aspects of their validity such as precision, specificity and reproducibility. This problem has been largely compounded by the fact that techniques, especially those used in biomedical studies have become much more sophisticated and complicated. As a result, many commercial companies have produced kits containing all the reagents needed to perform a given technique. Thus, many kits have appeared which, associated with simple instructions for use, make it possible to carry out, for example, the determination of hormones or other metabolites. All this creates an impression of great ease but, if we are not careful, this can also potentially lead to the production of invalid data.

This problem has, for example, become very present in endocrinology following the generalization of steroid hormone assays by radioimmunoassay (RIA) or enzymo-immunoassay (EIA). The relative ease of use and exquisite sensitivity of these assays, which in the best cases allow the detection of a few picograms (10^{-12} grams) of steroid, opened the door in the 1970s to the longitudinal study of hormone concentrations (precise hormone monitoring) both in humans and in small laboratory animals such as rodents (rats and mice). However, these assays suffer from a potential weakness in their specificity: compounds, similar or not to the steroid under study, can interfere with the quantification and lead to artificially high results. When these techniques first appeared in the 1970s, scientific journals generally required that any study using a new RIA to determine a steroid in a new species include extensive validation of the procedure. It is now quite common to use an assay kit validated for humans to assay steroids in various animal species and validations of this novel use are not always presented (Balthazart 2020). Similarly, the visualization of various compounds within biological tissues by the technique of immunocytochemistry has become common use but validation of the specificity of the immunological detection is not always performed. Several journals have become alarmed by this situation and have published specific instructions on this subject (Saper and Sawchenko 2003; Saper 2005; Rhodes and Trimmer 2006; Gore 2013). The results of such technically flawed work will obviously not be replicable in many cases.

7.1.4. *The low power of the studies*

The researcher who wishes to publish quickly will also be tempted to use a limited number of animals or experimental points to draw conclusions. These conclusions will be less reproducible if the number of experimental points is small. We might naively think that if a statistically significant difference is detected, its importance is greater if it can be demonstrated with a small sample size. However, methodologists have shown that the exact opposite is most often true.

This phenomenon has been called the *winner's curse* (Button et al. 2013). It is explained by the fact that if a study shows an effect based on a small sample, there is a good chance that the observed difference is the result of fluctuations in the sampling favorable to the thesis being studied. Indeed, if the actual size of a difference between two populations of data is small, the difference observed in a small sample is more likely to be detected if the sampling is non-representative and artificially increases the size of the difference. This inflation is worse for small samples and in practice means that the smaller the sample, the greater the probability that the observed difference is the result of sampling fluctuations.

Much biomedical research is in fact devoted to the study of differences that have a limited magnitude. Mathematical simulations too complex to be discussed here indicate consequently that a rather large fraction of the observed and thus published results (from 10 to 80% depending on the size of the actual difference and the sample size) are non-reproducible and suffer from this *winner's curse* (Button et al. 2013). This is probably one of the main causes of the non-reproducibility problem that plagues science today.

7.1.5. *Misuse of statistics and pseudoreplication*

REMARK.– This section is presented in an extremely simplified manner so that it can be understood by as wide an audience as possible. I am well aware that the wording may potentially offend specialists.

Any scientific study today must base its conclusion on a statistical analysis of the results. This makes it possible to avoid the influence of the cognitive biases described in section 7.1.1 and also to base any decision on standardized procedures that will be the same for an American, European, Chinese or Japanese researcher. This is one of the conditions that allow scientific conclusions to be validated and acceptable worldwide.

This analysis, if properly conducted, indicates the likelihood that an observed difference is the result of random fluctuations in the sample rather than the demonstration of a real difference. It is thus measuring the risk of concluding that there is a difference when there is none. Classically, it is considered that if this probability is less than 5% ($p < 0.05$), then there is a real difference. This probability is calculated by taking into account the number of experimental data points, the magnitude of the observed difference between groups and the variability present in the data. It is directly proportional to the first two of these factors but inversely proportional to the second. For this probability to be correct, it must, of course, be calculated by fulfilling a whole series of conditions which are again too complex to be discussed here but are discussed in detail in many books on methodology or statistics specifically dedicated to this subject.

For example, experimental data submitted to statistical analysis must meet a series of very specific conditions. In order to properly use the most common statistical techniques, i.e. the so-called parametric statistics such as the Student's *t* test comparing two averages or the analysis of variance which allows several averages to be compared, the data must in particular be normally distributed (according to the so-called Gauss curve) and have non-different variances. Parametric analysis of data that deviate significantly from these conditions will lead to erroneous results, either identifying differences that do not exist or failing to identify differences that are present. Many published studies, however, violate these very basic conditions. There are, however, relatively simple ways of analyzing data that do not meet these conditions. These are the so-called non-parametric statistical techniques in which the initial data are ranked in ascending order and the analyses are then carried out on the ranks rather than on the raw data. The Student's *t* test is replaced by the Mann Whitney test and the analysis of variance by a non-parametric analysis of variance known as Kruskal–Wallis. Awareness of this problem has developed over the last two decades, but we still encounter too many publications that are affected by this very obvious problem.

Another fairly recurrent problem concerns the presence of pseudoreplication in the data (Lazic 2010). The data that are submitted to the analysis must be statistically independent. In other words, the different experimental units that are compared by the statistical test must not be related to each other and must not be able to influence each other in any way. This may seem a little abstract, but a few examples will help to illustrate the point.

If researchers are interested in the influence of an experimental treatment on the size of neurons in a specific region of the brain, they can measure this size on 50 neurons in an experimental individual (A) and on 50 neurons in a control subject (B). A statistical test could show a statistically significant difference between these two groups of data, but this test would not tell anything about the existence or not of a treatment effect. The two groups of 50 measures are indeed interdependent, and it is quite possible that subjects A and B differed before the treatment was applied. The statistical test must be performed on independent experimental units in order to assess the real probability that the treatment had an effect. A test such as the one carried out here indicates that A and B are indeed different but says nothing about the treatment because the number of independent observations is 1 + 1 and not 50 + 50. It is essential to multiply the number of subjects exposed to the treatment and its control in order to draw reliable conclusions.

The same problem is present when we subject pregnant mice to a treatment and its control and study the impact of the manipulations on the pups. The number of independent subjects here cannot be the number of pups but the number of pregnant mothers. If, for example, we have three treated mothers and three control mothers who have each given birth to eight pups, the statistical test that could show a treatment effect must be on 3 + 3 mothers and not on 24 + 24 (3 × 8) offspring.

This problem of pseudoreplication is in fact very insidious. It occurs to some extent if we compare the response to two different stimuli (e.g. the song of an individual of the same species and the song of another species) in five control subjects and five subjects undergoing an experimental treatment. If the stimuli used are the same for all subjects, the conclusions will relate to the two stimuli used and cannot be generalized to other stimuli of the same type.

Similarly, if the five control subjects and the five experimental subjects are kept in two cages, one per group, they are no longer totally independent. They could influence each other and thus induce a difference between groups that would not be related to the experimental manipulation being tested. For example, a sick or a very aggressive mouse in a cage could modify all results of this cage. Any variable shared by all subjects in the control and experimental groups thus weakens the certainty that we can have about the effect of the experimental treatment. This is true for the same syringe used to inject all subjects in the same group (it could be contaminated by bacteria) or even for a preparation used during the treatments that is the same for all subjects in a group. This type of problem is in fact difficult, if not impossible, to avoid completely. If two groups of animals are tested for the effect of a particular drug or hormone compared to a control, there will often be only one preparation of the test product. If, as a result of a problem in the chemical synthesis, this preparation does not actually contain the active ingredient or contains a contaminant molecule that has an effect by itself, the interpretation of the experiment will be distorted.

The most robust solution would, of course, be to use an independent preparation for each test subject, but this is in most cases impossible. We must therefore always be aware of this limitation and carefully evaluate to what extent any form of interdependence existing between subjects in a group could affect the final result.

7.1.6. *Probability hacking (P-hacking)*

Computing statistical tests has been greatly facilitated since the appearance and generalization of personal computers. In 1970, comparing two groups of data by a simple Student's *t* test implied calculating the mean and variance of each series of data and then calculating the test itself, which was a long and tedious task (see Chapter 1). This work could initially only be done with the help of large computers where the data were introduced using punched cards. Such work can now be done in a few seconds on personal computers that allow for rapid analyses of the same dataset using various statistical techniques (parametric or non-parametric analyses, use of various transformations of the data, analysis of variance, regression, principal component analysis, etc.) with emphasis on one or another aspect of the results. The temptation is therefore great to carry out multiple analyses and to present in the final publication only those that indicate the existence of significant effects. This phenomenon has been called "probability hacking" (P-hacking), and, of course, it distorts the final picture that emerges from a study.

A statistical test assesses the probability that a difference between groups is the result of the experimental factor being tested rather than fluctuations in sampling. The standard way to accept the existence of an effect is to have a probability of less than 5%. The repetition of tests poses a problem that has been well identified and resolved in the specific case of comparisons of several experimental groups studied in parallel. In a simplified way, if we analyze the results of an experiment comparing seven groups of subjects, in order to compare these groups two by two by Student's *t* tests, we will carry out 6 + 5 + 4 + 3 + 2 + 1, i.e. 21 tests (N × (N-1)/2) and one of them should be significant at the 5% level (1 out of 20) by simple random fluctuations in the sample in the absence of real differences between the populations (21 X 5% >100%). Appropriate statistical techniques (specific post-hoc tests following an analysis of variance identifying the existence of an overall effect at the level of the seven populations) have been developed to take account of these multiple tests, but they are not always applied. Similarly, if we study, for example, the potential effect of various endocrine disruptors on about 50 dependent variables (e.g. body mass, fertility, levels of a dozen hormones, aspects of brain structure), a significant difference will always be found in the dataset, but it is doubtful that it will be reproducible if it is not associated with a potential mechanism of action of the disruptors.

The same problem is present if we multiply analyses of the same dataset until one of them is significant and only that one is reported. This problem has been highlighted in particular in nuclear magnetic resonance (NMR) studies of brain activity quantifying the level of blood deoxygenation (BOLD). It is possible to analyze this type of data by a multiplicity of image processing and statistical calculation methods. One publication even showed that by applying the 7,000 analysis methods that can be found in the literature to the same dataset, we end up finding a significant effect 90% of the time for at least one brain area in a dataset where there is no real effect (Carp 2012). The reproducibility of this kind of data is therefore very low.

7.1.7. *Inadequate presentation of results*

Finally, it should be noted that for a study to be truly reproducible, the article describing it must present its methods in a sufficiently detailed manner to allow another researcher to reproduce it identically. Many articles published today are clearly deficient in this respect. Very critical information is often missing. This may include the sex of the subjects enrolled, their age, their weight or their previous experience. All of these variables can have a major influence on the results that are obtained.

This is particularly true for the sex of experimental subjects. Sex differences are nearly ubiquitous in all areas of biology and have multiple causes that have been reviewed many times. They are, for example, described succinctly in a recent review published by the Endocrine Society (Bhargava et al. 2021). These causes fall into four categories: (1) genes (n = ± 100) present on the Y chromosome that produce male-specific effects, (2) a dosage effect of genes located on the X chromosome related to incomplete inactivation of one of the two Xs (± 350 genes escape this in activation) or parental impregnation of X, (3) differential early hormonal impregnation in the two sexes and (4) differential experiences during ontogeny. The combination of these mechanisms thus produces two major types of organisms, males and females, which differ in a large number of characteristics.

Sex differences affect a wide variety of biological processes. Almost all are impacted so that the default hypothesis must, at this point, be that any phenomenon studied is sexually differentiated rather than the other way around.

It should be noted, however, that the magnitude of these sex differences is variable depending on the type of phenomenon studied: it is in rare cases qualitative (sexual structures) but more often quantitative with an observed standard effect size (ratio of the absolute difference between sexes to the common variance) that can be small (d = 0.1 - 0.2) or large (d = 1 or 2).

For example, sex differences have been identified in: overall brain size and structure, connectivity and function studied by BOLD-RMN, brain–stomach relationships, predisposition and control mechanisms of obesity, responses to stress, and to its pharmacological treatment, physiology of the cardiovascular and renal system, pain perception (nociception), drug addiction and pharmacokinetics in general (Bhargava et al. (2021) for more information).

Ignoring these differences can induce a major experimental bias that can weaken the methodological quality of the experiments concerned (Karp and Reavey 2019). For example, working with subjects of unknown sex can lead to increased within-group variance, and thus a decrease in the ability of experiments to detect a true difference between experimental groups. Alternatively, this can induce statistically significant differences that do not actually exist, if subjects of a given sex accumulate in one group. These potential pitfalls are all the more to be feared as the numbers of subjects in the experiment are relatively small (and the statistical power is therefore low, see section 7.1.4). Moreover, a variation in the sex ratio from one experiment to another strongly reduces the chances of replicating the results. Finally, results obtained from experiments using only one sex cannot logically be extrapolated to the other sex without major limitations, if at all.

Historically, preclinical and clinical studies have largely been conducted in men, primarily for safety reasons (Holdcroft 2007b) (risk of testing a drug or procedure on a woman who might unknowingly be pregnant) (Holdcroft 2007a). It was somewhat naively assumed, before the extent of sexual differences was so clearly identified, that the results would be applicable to both sexes. This has led to many clinical problems. For example, the same dose of aspirin has different effects in men and women, and a drug such as zolpidem used to treat insomnia should be used at different doses in the two sexes (Sastre 2014). In response, the National Institutes of Health (NIH, Bethesda MD, USA) issued in 2014 a series of recommendations on this topic and decided that they would no longer fund research that was not based on the study of both sexes, except for duly justified exceptions (e.g. study of prostate or uterine cancer) (Clayton and Collins 2014; Clayton 2016). Note on this occasion that these recommendations also apply to work done on cell cultures: the sex of the organisms from which the cells originate must be taken into consideration because it is well established in many fields that sex influences the physiology of cultures (Clayton 2016; Rich-Edwards et al. 2018).

Any experimental study should therefore at a minimum (1) account for the sex of subjects in the experimental design (as an independent variable), (2) collect data by separating those for males and females, (3) analyze them accordingly and (4) present them separately in reports and publications (see Diester et al. (2019) for a detailed description of this NIH-recommended 4-C approach of considering, collecting, characterizing and communicating the information). Such a study will not necessarily identify a sex difference with sufficient power (conventionally set at an acceptable minimum of 80%), but it will most importantly ensure that the results obtained are not in a major way biased by an unsuspected effect of sex. This will help guide future experiments (Diester et al. 2019). Currently, the lack of information regarding the sex of subjects in a publication is, however, a major problem that greatly affects its replication.

Furthermore, the reagents used in an experiment should be completely identified. We have already mentioned the fact that antibodies used for immunocytochemical visualization of a molecule in histological sections may cross-react and therefore give false positives. This lack of specificity can vary between antibodies directed against the same antigen but prepared and supplied by different companies. Moreover, in the case of polyclonal antibodies, this specificity will potentially be different from one batch of antibodies to another because they represent either different bleeds from the same donor (rabbit, goat) or samples taken from different animals. The full reference of an antibody used, including the manufacturer and the batch number, is therefore a critical information that should always be mentioned.

The results must also be presented in sufficient detail so that their nature can be clearly understood. Ideally, researchers are now asked to present the individual points in the data and their distribution. The same average statistical results can indeed be obtained from very differently distributed data as elegantly described by the Anscombe's quartet (see Figure 7.2).

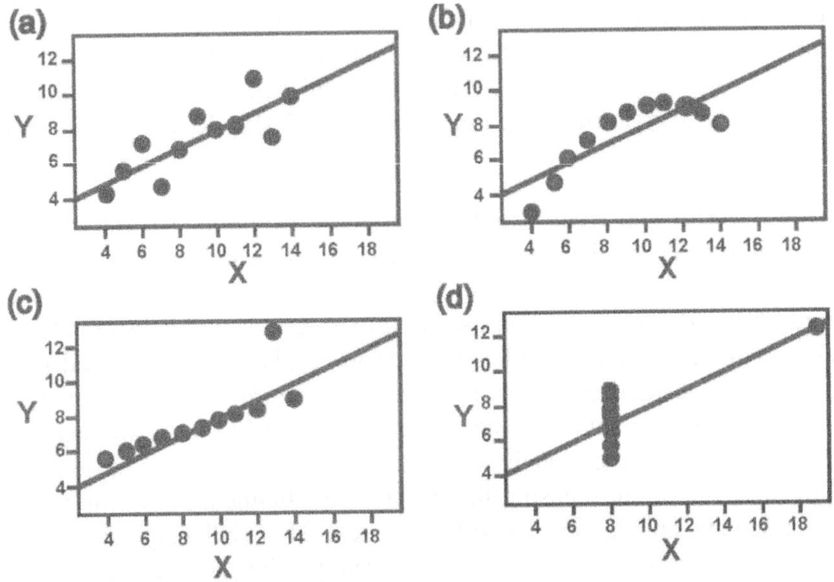

Mean of X	9.00
Variance of X	11.00
Mean of Y	7.50
Variance of Y	4.125
Correlation between X and Y	0.816
Regression of X on Y	Y=3.00+0.500 X

Figure 7.2. *Anscombe's quartet presents four sets of figures illustrating a hypothetical relationship between two variables X and Y. For a color version of this figure, see www.iste.co.uk/balthazart/fiftyyears.zip*

COMMENT ON FIGURE 7.2.– *These four series have the same mean, variance and linear correlation, but the individual points make it clear that these series are very different*[2].

2 Redrawn from the Wikipedia site dedicated to this problem (see https://en.wikipedia.org/wiki/Anscombe's_quartet).

All these problems described so far are unfortunately present in a significant way in the literature, even though they generally fall under the heading of honest error, i.e. made without any intention of fraud. The same is not true for the problems discussed in the next section.

7.2. Scientific fraud

The pressure on researchers to publish more and more in journals with a higher impact factor has unfortunately led a small number of researchers to violate commonly accepted ethical rules. These violations are of three distinct kinds.

7.2.1. *Data selection*

A paper that is likely to be accepted and published in a high-impact journal must tell a new story on a subject of fundamental interest and also present a series of experimental results that solidly support the story presented. It is rare that in a series of experiments, all results are in perfect agreement with the thesis presented. In my 50-year career, which has resulted in more than 500 publications, this has only happened a few times that can be counted on the fingers of one hand. There are always experimental points, or groups of subjects, or even entire experiments that seem to contradict the proposed thesis or at least do not provide arguments to defend it. The temptation is therefore great to simply ignore these annoying data in the publication.

There are codified rules that allow the removal of a few rare experimental points either if they are the result of an obvious technical error or if they appear statistically as outliers in the study population. This removal of data points must be clearly stated in a publication, and these deletions of data must be limited in number. However, the line between legitimate selection of the few data to be discarded and manipulation of the results by omission is often blurred.

Given the number of papers published in high-impact journals that present "perfect" sequences of data, I am personally tempted to think that data selection has become quite widespread, but it is difficult to really quantify the phenomenon. The extent of suppression is obviously critical. If a researcher performs five experiments, four of which show the same thing and the fifth, not reported, is inconclusive, it is likely that the published result is correct and reproducible. However, if we are in the opposite situation where only one of the five experiments gives the expected result and that is the one that is published, this work most likely leads to a dead end. There is a gradient here from responsible and perfectly acceptable data processing to outright lying by omission. This is not the case for the following category of problems, which are always outright fraud.

7.2.2. *Partial or total data fabrication*

Unfortunately, over the last few decades, a limited but worrying number of scientific publications have been identified in which all or some of the data were simply invented. The motivations that drive a researcher to such misconduct are complex, but often have to do with the search for recognition. The mystification can sometimes last for years, especially in less active research fields where replication of a result will not be attempted until years later, but it is often discovered for a variety of reasons discussed in the next section.

In a very active field such as biomedical research, this type of fraud often lasts only for a short time. An example is the article published on May 22, 2020 in the prestigious medical journal *The Lancet* describing a study based on a very large number of cases (96,000 patients in 671 hospitals) that supposedly showed that the treatment of Covid-19 with hydroxychloroquine did not have a beneficial effect and was even associated with higher mortality and heart problems. It soon became apparent that this study was based on data that would have been impossible to obtain in the time elapsed since the pandemic began. Additional inquiries confirmed these doubts, and as a result, the study was retracted on June 4, after less than two weeks. This is, of course, an extreme case where research on the topic covered in the publication was unusually active.

The prototypical example of fraud identification after a much longer period of time concerns the British psychologist Cyril Burt (1883–1971), who throughout his life published studies aimed at demonstrating the mainly hereditary character of intelligence. These studies were based on the analysis of the intelligence quotient (IQ) of identical twins (monozygotic) separated at birth and raised in different environments. Burt attempted to demonstrate that the IQs of identical twins remained highly correlated despite their different life experiences and learning opportunities. He published numerous articles supporting this thesis. Shortly after his death, doubts began to be raised about the nature of the data contained in his publications. Suspicion initially developed because Burt had published three studies based on different samples but showing a correlation identical to the third decimal place (0.711), which seemed highly improbable.

Historical studies later showed that one of the collaborators who had supposedly helped Burt collect these data never existed. After explorations, most of this researcher's work was discredited as tainted by problems of manipulation and falsification of results that were apparently caused by a paranoia of the researcher who wanted to prove his thesis at all costs even against the results obtained[3]. This attests to the fact that the problem of scientific fraud is not new and is not only caused by the pressure to publish, even though its incidence has probably tended to increase following the aggravation of this pressure in the scientific world.

Although the critical evaluation of published research has become increasingly easier with the availability of digitized publications, we are entitled to wonder whether the cases identified are not just the tip of the iceberg that would have a much greater importance. The sanctions taken against the researcher convicted of such practices are obviously very severe, ranging from the banning of all forms of grants to the exclusion of the researchers from their own university.

7.2.3. *Plagiarism*

This last type of misconduct involves copying all or part of the work of another author or in attributing to oneself the ideas of another researcher without citing their sources. It is a form of theft that causes moral or

3 https://en.wikipedia.org/wiki/Cyril_Burt.

commercial damage to the original author if financial interests are at stake (e.g. related to the commercialization of a medical treatment). The publisher who owns the reproduction rights of the original article is also harmed. Plagiarism is subject to criminal sanctions and is, of course, monitored by all journal publishers, research funding agencies and universities themselves. It should be noted, however, that in its simplest version plagiarism does not generate erroneous scientific data and therefore does not jeopardize the reproducibility of science.

A particularly interesting case concerns self-plagiarism, i.e. the republication by an author of a text that has already been published and for which he/she has transferred the copyright to the publishing house concerned. It is, of course, a question of degree. We can easily agree with the idea that republishing in extenso an article that one has already published, whether in the same language or in a different language, is completely unacceptable and should lead to sanctions if the fraud is discovered. This being said, a researcher is often led to publish several articles on the same theme, describing progressive advances on the same question and using all or part of similar methodologies. The introduction to all these articles should therefore describe the state of the question, and they must include a methods section that covers all or part of the same methods. Researchers and editors are split about whether it is permissible to reuse parts of the previously published text. It will often be possible to change the phrasing and structure of an introduction, but there are only limited numbers of ways to describe an assay or sample collection method. In my opinion, limited self-plagiarism should be allowed in these cases.

All problems described in this section 7.2 ultimately result in the publication of poor-quality research, which has two types of negative consequences. On the one hand, they obviously undermine the validity of the knowledge acquired and its credibility. This loss of confidence affects the researchers themselves and also the general public. The result is a distrust in science that is sometimes associated with the development of irrational beliefs (see Chapter 10). This mistrust may also affect the governments that decide on the allocation of research funds and, at the very least, temper their enthusiasm for granting subsidies to this enterprise, which is nonetheless crucial for our societies.

On the other hand, poorly conducted or fraudulent research wastes valuable resources including taxpayer funding, the lives of experimental animals or patient data. The publication of these data also commits other researchers to research paths that are often dead ends, leading to further waste of resources. There is thus damage to the society in general. The scientific community is therefore fighting proactively against this type of downward spiral, using several means.

8

The Scientific Community's Fight Against these Aberrations

The problems described in Chapter 7 affect the progress of science, but several safeguards have been put in place to limit their impact. The first and most important of these mechanisms is based on the fact that every article published in a serious scientific journal is first critically evaluated by a scientific editor and by two or three researchers working in the same field. This is called the peer-review system.

8.1. Peer review

Typically, a researcher who wishes to publish an article he or she has just written sends it (nowadays electronically) to the editor of a journal who will distribute the manuscript to two or three researchers with expertise in the field, giving them 10–30 days to evaluate it and send their comments. They will read the text in detail, assess the validity of the methods used, the statistical techniques on which the conclusions are based and the validity of the conclusions drawn from the data. They should also be able to see whether the article makes appropriate reference to previous work that has studied the problem.

On the basis of these comments, the editor will then decide whether the article has fundamental problems and should be rejected, or whether it is acceptable for publication as such, or, as is most often the case, whether the article is potentially acceptable but contains small problems that should first be corrected. These comments may concern a request for further explanation of aspects of the methods or results, a request for modification/correction of the statistical analyses or mode of presentation of the results, or a challenge to aspects of the conclusions that are not fully supported by the results. After correction by the author(s), the article will be resubmitted to the journal, again evaluated either by the editor alone or by the editor and the original reviewers if the changes made are significant. If all the requested changes have been satisfactorily incorporated, the article will finally be accepted for publication.

In addition to this evaluation of the quality of the work, a check for plagiarism will often be performed. On the one hand, the two or three reviewers will often be familiar with the literature in the field and should be able to detect whether the data presented have already been published in whole or in part. In the worst cases of plagiarism, such as copying an entire text or parts of it, they might also be able to see whether parts of the text are duplicates of earlier texts, but this requires an excellent memory for detail that is not always present. However, the digitization of all previously published work has made it possible for some years now to develop plagiarism detection software that can compare in detail any text submitted for publication with all previously published texts that are available in digital form. This is a colossal database, as most publishers have now digitized all their journals published in the last 20–30 years, or even since the birth of the journal (see section 3.1).

I have access, as associate editor of the journal *Hormones and Behavior*, to such software developed by the Elsevier publishing house that publishes the journal. This software, called Similarity check/iThenticate, is actually amazing. It provides a detailed report containing an overall assessment of similarity with any previously published text, and it also identifies and reports separately any group of a few words that are identical to a published group with the precise reference of the text concerned. Not all identity is necessarily plagiarism, of course. Thus, the software indicates a match for a text as short as “This finding is consistent with studies that have…”. Clearly, there is nothing objectionable here, because there are not a thousand ways to say one thing. However, an overall percentage of similarity and total number of identical words to a previous publication are provided. No automatic decision can be made from this report, but a score above 20–30% similarity will attract the attention of the editor who can then analyze the problem in depth. Human experience is therefore still needed to reach a final decision and possibly to retaliate against the author. The similarity could indeed refer to a pre-publication in a site such as BioRXiv, which is authorized by most journals.

This whole evaluation process of course limits the presence in the published literature of the problems described in the previous section, but this evaluation is a human activity and, as such, is sometimes subject to failure. It should also be noted that many journals have over the past decade greatly increased their requirements for providing detailed information about the results that form the basis of a paper submitted for publication. For example, many journals now require that all of the individual data (point by point) that form the basis of the findings be published in parallel on a specialized website (such as Dryad)[1] so that the statistical analyses can be reproduced by any interested colleague. In addition, many journals have added requirements regarding the presentation of results and information regarding method validations. For example, some journals require that histograms depicting the means and variances of different datasets be augmented by the illustration of all individual points (see the example in Figure 8.1). This of course makes it more difficult to manipulate the data and makes it easier to detect possible manipulations.

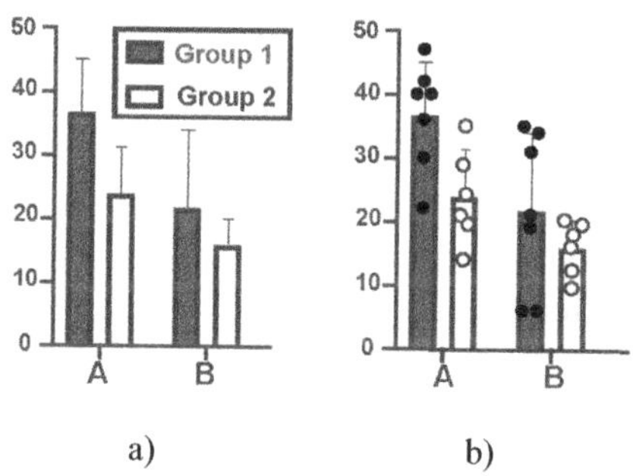

Figure 8.1. *A) "Classic" presentation of a result histogram and B) its now often required version, including individual points. For a color version of this figure, see www.iste.co.uk/balthazart/fiftyyears.zip*

8.2. Post-publication criticism by the entire scientific community

Even though poor quality or even fraudulent data make it through these filters and can be published, they will always be subject to critical evaluation by the scientific community.

1 www.datadryad.org/.

A fraudulent publication can thus be identified very quickly (see the case of the *Lancet* article in section 7.2.2), but in other cases the deception may last for years, particularly if it concerns a subject that does not have a direct societal sequence (e.g. in health or economics) or a field of science that evolves more slowly, and therefore where an attempt to replicate or analyze a phenomenon in greater depth may not be undertaken for years. It should be noted in passing that these are areas where the damage created by fraud is not as great because it initially concerns only theoretical knowledge.

The evaluation of published research is therefore an ongoing process and the disclosure of problems in published work is now facilitated by the existence of specialized websites. Thus, the PubPeer platform[2] collects comments mainly from anonymous researchers about articles after their publication and indicates potential problems or ethical breaches. The nature of these problems varies and ranges from the discovery of factual errors in statistical analyses to the manipulation of images representing histological structures or electrophoresis (e.g. Western blots). Image manipulation has now become very easy with image processing programs such as Photoshop™, but detection of such manipulations is also simplified by other software.

Many problems have been identified. The PubPeer contained no less than 120,000 comments on 40,000 scientific publications by 2021. This initiative has received a mixed reception. Some see it as a system of generalized denunciation, while others consider it a necessary evil given the difficulties of obtaining a more effective system of self-correction. It should be noted that, although the anonymity of contributors is protected, there are moderators of the site who only admit comments that are related to objective and verifiable facts.

8.3. Withdrawal of erroneous or fraudulent items

The accumulation in PubPeer of suspicious comments about a researcher's publications often leads to the opening of an internal investigation in the employing institution. An investigation can also be triggered for other reasons: research by colleagues who fail to reproduce a result, denunciation of data manipulation by the co-author of an article and

2 www.pubpeer.com.

identification by any researcher of an incongruity or practical impossibility in the exposed research.

If such an investigation is initiated, often by the university where the researcher works or by an organization that funds the researcher, all original documents underlying the publications will have to be produced and will be examined in detail. Depending on the outcome of this investigation, sanctions may or may not be taken and the article(s) in question will be retracted. Another specialized site lists all these retracted articles[3] and the newspaper where they were published will mention it in a note. However, this information is often not widely known and it is sometimes observed that retracted articles continue to be cited in the literature, whether in good or bad faith.

A recent article published in the journal *Science* analyzed this problem and regretted that retracted articles are still quite abundantly cited (Brainard 2022). He cites the example of the scientific output of a now deceased Japanese nutrition researcher, Dr Yoshihiro Sato, who turned out in retrospect to be one of the most prolific fraudsters known in science. A total of 27 papers by this author have been officially retracted. A systematic review of the literature identified 88 review articles in the biomedical field citing these 27 more than problematic articles between 2003 and 2020.

According to Brainard, the conclusions of 39 of these 88 syntheses would be significantly weakened and altered if the retracted articles were excluded. More disturbingly, the authors of 86 of the 88 articles in question and most of the publishers who had published the summaries were contacted by the investigator, but after a year the journals had published warning notices for only eight of the articles. In addition, they received no response from half of the authors, and the others replied that they would not change their reviews because they were either too old or because they did not have the time. Brainard calls these retracted articles zombie articles: even after retraction, which should constitute a death sentence, they live on through citations by people who may or may not be aware of their status.

This phenomenon is unfortunately not an isolated one, as noted by Ivan Oransky, co-editor of Retraction Watch. For example, a 1998 study in the journal *JAMA* showed that 94% of the 299 articles retracted in the field

3 www.retractionwatch.com/.

were still listed without comment in the MEDLINE database. The continued existence of retracted articles sometimes has important practical consequences.

For example, although the article has been notoriously retracted, many vaccine opponents continue to rely on an article by Andrew Wakefield published in *The Lancet* that allegedly showed a link between the MMR (measles–mumps–rubella) vaccine and autism. The article was retracted in 1998 because the author had acted dishonestly and misrepresented his results. Dr Wakefield had also failed to declare that he had received grants from lawyers working on a potential lawsuit against vaccine manufacturers (a known conflict of interest). Furthermore, several studies have long since demonstrated the absence of any link, but the article is still used as an argument by the anti-vax community[4].

Similarly, we continue to see regular references to the article by Jacques Benveniste published in the journal *Nature*, which claimed to show the existence of a memory of water, i.e. the capacity of the liquid to keep a trace of a compound that has been dissolved in it, and this even after impressive dilutions that make that there is no longer a single molecule of the compound per liter of water (Davenas et al. 1988). A comment by the editors of the journal indicated already at the time of publication that many experts considered this result incredible and probably impossible. This idea was subsequently subjected to replication attempts by a variety of experts including scientific researchers and a former magician who was better able to discover hidden artifacts. The conclusion of these studies was that the "water memory" effect is impossible to replicate and probably the result of a technical artifact[5]. Although the paper was never formally retracted, very clear comments casting doubt on it were published. It has also been shown that there was also an obvious conflict of interest: Dr Benveniste's research was financed in part by the Boiron firm, the leader in France in the field of homeopathy. This is a parascientific or pseudoscientific technique based on the use of dilutions similar to those described in the article by Benveniste's group in *Nature*. Although no support for this theory is available to date, it continues to be cited in writings by proponents of homeopathy[6].

---

4 https://en.wikipedia.org/wiki/MMR_vaccine_and_autism.

5 www.en.wikipedia.org/wiki/Benveniste_affair.

6 https://en.wikipedia.org/wiki/Water_memory.

The scientific community is beginning to take this problem very seriously, and some bibliographic databases such as EndNote™, Papers™ or Zotero™ now prominently indicate whether an article has been picked up in the Retraction Watch database. Some researchers have also decided to devote a substantial portion of their time to tracking down malfeasance and fraud in scientific publications. This is the case of the Dutch microbiologist Elisabeth Bik, who now specializes in detecting images that have been manipulated in violation of ethical rules[7]. In this context, she works with the PubPeer and Retraction Watch platforms. She has identified more than 2,000 articles containing suspicious images and has identified a Chinese company responsible between 2018 and 2020 for 400 different articles containing the same fraudulent images. These articles are then sold to Chinese medical students. These students are required to publish at least one article in an international scientific journal in order to obtain their final degree, but their curriculum does not give them the opportunity to produce original research. It is of course disturbing to realize that these falsifications were not identified during the evaluation process of these articles. More recently, Elisabeth Bik published in PubPeer about 60 comments pointing out various unethical actions (anomalies in results, conflicts of interest, ethical problems related to the use of human patients) in the publications of Dr Didier Raoult and his colleague Éric Chabrière of the Institut hospitalo-universitaire Méditerranée Infection. These researchers have filed a complaint for moral harassment and the case is now in the hands of the courts. This obviously shows one of the problems associated with this system of correction of scientific literature.

7 https://en.wikipedia.org/wiki/Elisabeth_Bik.

9

Essential Modifications

From what has been said so far, it seems clear that a fairly profound modification of the way researchers are evaluated is needed. Paraphrasing Marcus Munafo and his colleagues (Munafo et al. 2017), we can indeed say that in the current system, “researchers tending to maximize their productivity should devote most of their efforts to conducting small studies with low statistical power”, and we could add: without too much worrying about the quality of their work. This leads to bad science, but it is obviously within the ethical rules.

As described in Chapter 7, the pressure to publish can also lead to real fraud, ranging from data selection to outright fabrication. The first problem is, however, much more widespread than the second. The selective publication of results seems to be the most frequent problem that threatens the effectiveness, validity and therefore the reproducibility of science.

The solution to this problem requires both a change in the process of evaluation and publication of results and a complete reorganization of the system of rewards for researcher promotion or grants distribution. This requires the involvement of both scientific journals and evaluation bodies, funding agencies and universities themselves, in order to induce and encourage a change in the behavior of researchers.

9.1. The publication process and peer review

First, it is necessary to change the form of the published articles to better document the results presented in them. This point has already been discussed in a previous section devoted to addressing the identified problems. Overall, this increase in transparency in the presentation of results includes a (more) detailed description of methods and their validation, the presentation of individual data points in figures, and the adoption of detailed well-described and validated statistical methods of analysis. It could or should also require the deposit of raw data on the specialized websites that now exist. All of these measures have in fact already been implemented by a number of journals.

It would also be necessary to increase the rigor of peer review. This should be facilitated by the adoption of the measures described in the previous section, but this task is made more and more difficult by the quantity of articles currently submitted for publication. A staggering number of scientific articles are published each year and their evaluation takes up a significant amount of the researchers' time.

For example, it is estimated that the critical appraisal of manuscripts published in 2020 consumed a total work time of US scientists equivalent to $1.5 million[1]. Consequently, it is becoming increasingly difficult to find reviewers as indicated by my personal experience and that of my close colleagues. As an associate editor of the journal *Hormones and Behavior* (Elsevier), I am responsible for organizing the review of more than 50 manuscripts per year. Finding reviewers is becoming increasingly difficult, and it is not uncommon to have to contact more than 10 different researchers before finding two who agree to do the work.

It is also necessary to campaign to make authors and potential reviewers aware of their responsibilities. Considering that in an average journal, 25–50% of submitted articles are accepted for publication, and that each submitted article must be reviewed by at least two experts, every researcher should therefore consider himself or herself morally obligated to review four

1 *Science*, vol. 374/6571, p. 1033.

to eight articles for each article he or she publishes. Some reviewers are dedicated to this task, but others are much more selfish and review few or no manuscripts in their field of research, despite requests from publishers. This work is of course complex and time consuming. It should be taken into account in the evaluation of the researchers' activities. The tools to do this are already in place. The site Publons[2] has existed since 2012 and lists all the evaluation work carried out by researchers who wish to do so (over 500,000 in 2018). The evaluation is automatically communicated by many journals and for others, the researcher only needs to send Publons a copy of the journal's e-mail thanking them for their work and it will be automatically counted.

The site also provides statistics on the researcher's activity as an author and reviewer. As an example, I admit that I am still far from the goal described above since Publons tells me that during the last 12 months I have reviewed 2.2 articles (n = 16) for each article I have published. As an excuse, my activity as an associate editor additionally leads me to evaluate about 50 articles per year that are not counted. It should also be noted that, comparatively speaking, this insufficient reviewing score nevertheless ranks me at the 97th percentile of all researchers included by this site, and that the average researcher only evaluates according to Publons 0.3 articles per published article (Figure 9.1).

Figure 9.1. *Example of my statistics as an evaluator* ***published*** *by Publons. I have evaluated 89 articles in total, 16 of which in the past year (2021), which leads to 2.2 evaluations per published article*

The publication management site, maintained by Elsevier for the journal *Hormones and Behavior*, also compiles all invitations to review an article sent to each researcher and the number of times he or she has accepted or

2 www.publons.com, recently taken over by Clarivate and accessible at: www.access.clarivate.com/login?app=wos.

declined. It is thus easy to know each researcher's degree of involvement in the process. I know of close colleagues who almost never accept these invitations but continue to publish actively. So there is work to be done, because most of the evaluation is based on the work of a few. This information should certainly be taken into account in the evaluation of researchers and this activity should be rewarded.

9.2. Pre-registration of studies

It was also recommended that scientific research be systematically pre-recorded. Practically, the researcher who plans to test a working hypothesis should record in detail and officially (on a dedicated website) the protocol he or she intends to implement, including the methods by which the measurements will be collected, the number of subjects that will be used or data points that will be collected and the statistical methods that will be used for the analyses. This procedure would have several advantages, but also has certain disadvantages.

The most positive aspect is that this procedure, if generalized, would completely solve the problem of publication bias. It is indeed increasingly evident that few researchers spend their time trying to publish experiments that have produced negative results. Most journals, especially those with a high impact factor, are also reluctant to publish such data. As a result, the published literature contains only descriptions of studies that confirmed a basic hypothesis but not those that disproved it.

This problem is particularly prominent in the field of clinical experiments that study the effectiveness of a medical procedure or pharmacological treatment. These studies are themselves based on preclinical studies where potentially only the positive results have been published. If, in addition, only the positive clinical experiences are published and the negative ones are not, this could lead to the implementation of medical procedures with little or no effectiveness.

Various organizations, including the NIH, have expressed concern that there is a significant bias in the publication of negative and positive results of these studies (Johnson and Dickersin 2007; Hopewell et al. 2009). This is obviously a problem given the very significant waste of resources involved. This could also have negative consequences for a large number of patients if such a procedure is adopted when it does not have the expected efficacy. Consideration of all preclinical and clinical studies, both positive and negative, would obviously go some way to avoid this waste of resources and risk to potential patients.

Pre-registration of studies would also make it possible to combat the probability hacking described in section 7.1.6 more effectively since the statistical processing procedures of the data would be decided before they are collected. Similarly, it would become impossible to publish a study showing a given effect by claiming that it is the confirmation of a pre-existing hypothesis, when the study was initially intended to test another hypothesis and its purpose was changed after the results were obtained. This practice, referred to as HARKing (Hypothesizing After the Results are Known), has been identified as another bias in the scientific literature (Kerr 1998). Of course, a dataset showing an unexpected fact is worth publishing, but this should be presented as exploratory research and not as a hypothesis test, at the risk of distorting the nature of the experiment.

However, the pre-registration procedure creates an additional administrative burden and could, under certain conditions, represent a major limitation to creativity in research. At this stage, pre-registration has become a common practice in the field of clinical studies. It is still quite anecdotal in fundamental research. In many cases, it is not compatible with much of the basic research which is, at least in part, exploratory in nature, a feature that must of course be maintained. However, once an area of research is sufficiently mature and a piece of work sets out to test a specific hypothesis, the research should probably be pre-registered.

9.3. The reward system

In addition to changing the way research is conducted and published, changes in the way researchers are evaluated would, of course, also be required. The excessive focus on the number of articles published and their impact factor should be greatly reduced. Evaluation should be based on the quality of a researcher's output, assessed qualitatively by evaluating its originality, methodological quality, the amount of work behind the publication (to counteract the fragmentation of a dataset into multiple small articles of little significance), and the researcher's propensity to provide open access to his or her core data and to share them with colleagues to facilitate collaborations. The publication of negative data should also be valued, and this is as much a concern for journals as it is for granting agencies. It is currently impossible to apply such methods to the entire production of a researcher given the high number of articles that many publish each year It would, however, be possible, as some funding agencies already do, to ask a researcher applying for a promotion or a grant to list only the five publications in the field that he or she considers most representative of his or her production and to evaluate these few articles.

Other criteria could also be used, such as the researcher's contribution to the distribution of information in society via the publication of articles or books for the general public, also known as popularization, or his or her activity in evaluating articles submitted for publication (see section 9.1).

As far as the distribution of research grants is concerned, much more radical solutions have been proposed, such as the distribution of equal shares of the cake to all researchers, with the obligation to redistribute a fixed portion to colleagues who are considered to be deserving. Evaluation would thus be done on an ongoing basis by the entire academic community, and would no longer require the very tedious writing of grant applications, even though this activity is in part very creative. This option has already been discussed (see Chapter 6).

By bringing all of these incentives into play, we could hope, in association with the more thorough control of what is submitted for publication, to profoundly and sustainably change the behavior of researchers so that they focus more on quality than on quantity. All of this should ultimately improve the reproducibility of scientific research.

10

The Loss of Confidence in Science and the Return of the Irrational

Reading the preceding pages demonstrates in my mind that the profession of researcher has become infinitely more complex today than it was in the middle of the 20th century. At that time, science was still a largely curiosity-driven enterprise, and the financial aspects influencing the profession were relatively secondary. The researcher must now face working methods that are infinitely more elaborate and difficult to integrate and validate. Moreover, he or she is subjected to an increasingly complex administration and to a growing pressure to produce immediately usable data. This has led to drifts in the quality of work as we have explained in detail.

The research activity deployed by our societies in the Western world has also developed in an extremely important way. People are now employed in research who previously would have chosen another activity because being a researcher is not really for them a vocation motivated by a major desire to learn and discover. Therefore, we have a series of researchers who are "civil servants" of research and who work on this project from nine to five, five days a week. This has changed the landscape of universities quite drastically. Of course, there are still a certain number of extremely motivated people who will very often become leaders in their field, while the other researchers are in fact support staff for a line of research drawn by the minority.

10.1. A disaffection for science

The greater complexity that is present in the organization of research and in the results it produces also affects, of course, all other aspects of the functioning of our societies. Managing a country has become infinitely complex and is no longer understood by many citizens. The way science is perceived has also changed profoundly over the last few decades, and it is safe to say that science has lost much of the prestige it enjoyed in the middle of the 20th century. This loss of credit probably has several causes including the following:

– The current research is extremely complex, and it is increasingly difficult to explain it in a simple way to the public. I have myself carried out about 50 years of research work in the fields of ethology, endocrinology and neuroscience, but understanding an article presenting new results in cosmology, physics or chemistry is a very difficult if not impossible task for me. And even in the biomedical sciences, an article on cancer or immunology published in *Science* or *Nature* is incredibly complex to read. Only a simplified version explaining the meaning of the research is accessible. It is therefore not surprising that the general public has lost interest in the progress of this research, progress which is absolutely amazing in all fields. Just think of the two recent successes: the development of effective vaccines against Covid-19 in less than a year and the successful launch of the James Webb Telescope, an instrument that will allow us to explore the universe with a sensitivity never achieved before: it is capable of detecting in infrared the thermal signature of a bumblebee located at a stretch equivalent to the distance between the earth and the moon.

– The decrease in the quality of work (see Chapter 2). The reduction in the quality of work due to the intense pressure to produce (see Chapter 7) has undoubtedly also played a role. Many studies have shown that scientific work is now less reproducible, and this may have contributed to a loss of public confidence. However, this problem has been debated more within the scientific community itself than among the general public, and therefore is not, in my opinion, the *primum movens* of this disaffection.

– The general public has become aware, particularly during the Covid-19 pandemic, that the conclusions of scientific studies can sometimes change quite rapidly, especially if the research is intensive, as was the case with the pandemic. Some concluded that scientific results were not reliable. This is actually the result of a confusion that was previously widespread. What is unique about science is not that it proposes results that are established once and for ever, but that it presents results that are established with a precise and reproducible methodology. Science is not a series of results, it is a method. It allows for the establishment of models and theories that are then subjected to critical evaluation. If after many tests the theory continues to be accepted as the best explanation of the reality of the world, it is considered at this stage as established. However, it will always remain possible that subsequent work modifies its outlines, often marginally or more regularly by broadening and modulating the context in which it can be applied. This is the case of Newton's theory of universal attraction explaining gravity which was universally accepted for more than 200 years until the development of the theory of general relativity by Albert Einstein between 1907 and 1915. The latter has now integrated this theory of gravity in a much broader context. The discovery of this possibility of revision of scientific models has had the effect of disillusioning some people with science in general.

– It also became clear that science could not solve all the problems of society. In the 1950s and 1960s, it was thought that science was all-powerful and that technological advances would solve all of humanity's practical problems. Technology has indeed facilitated many aspects of everyday life and shown its power by sending people to the moon, but it has not solved the problems of poverty, social inequality and many others.

– Finally, in a very fundamental way, the second half of the 20th century was marked in the Western world by a progressive disinterest in the classically established religions, Christianity in the first place. This period was also marked by the decline of all the great economic models, communism, capitalism or liberalism, which clearly showed their flaws (the end of the "ism" theories). Communism collapsed first in most countries of the world, but liberalism and capitalism, which have now become untamed, are currently subject to intense criticism, and many have lost all illusions as to their capacity to manage the functioning of our societies in the best possible way. This questioning of most of the certainties that governed Europe for centuries has been extended to science, which is no longer considered the miracle solution to the problems of humanity.

This state of doubt towards science is unfortunately very regrettable because we live in a complex world where solutions to problems encountered should often come from an impartial scientific analysis. Just think about the pandemic that has been affecting the world since 2019 and the problem of global warming which is of enormous complexity but can (could?) be addressed with rational methods.

10.2. The development of irrational beliefs

This decline in "traditional certainties" has been associated with an increase in irrational beliefs and behavior; people have chosen other gods! The 20th century thus saw the development of a good number of pseudo-religious sects established by gurus self-justified by a so-called enlightenment but motivated most often by the lure of gain. Some of them established fortunes by exploiting the credulity of their flock. The public's loss of confidence in official science has also been the breeding ground for the development of alternative theories that claim to explain certain aspects of the world by models that are not based on any reproducible evidence and have even very often been shown to be incompatible with reality. A few examples will illustrate this point.

10.2.1. *Creationism and the denial of evolution*

In the United States, there are widespread movements that deny the theory of evolution and the evaluations of the age of the Earth and the universe generally established by cosmologists. These groups are often motivated by a literal reading of the biblical texts, which, as we know from the convergence of all studies of historical criticism, should be considered rather as pictorial narratives intended to illustrate general concepts and the means necessary for salvation without affirming anything at the scientific level. This interpretation was officially accepted by the Catholic Church at the end of the 19th century (see the encyclical *Proventissimus Deus* of Pope Leo XIII, promulgated in 1893).

It is now known that the form of these texts was deeply influenced by beliefs and traditions at the time when they were conceived and they should therefore not be opposed to the scientific data. Science indicates that the Earth was formed several billion years ago and not 6,000 years as deduced from a literal reading of the biblical texts. It is also clear that all animal species appeared through very progressive evolutionary mechanisms as proposed by Charles Darwin and confirmed with very precise mechanistic details by his numerous successors.

Nevertheless, very active movements denying evolution and trying to ban its teaching in schools persist in the United States. The statistics on this subject are in fact alarming since, in 2011, some 40% of Americans declared themselves to be creationists, while only 28% were in favor of evolution and 32% had no clear opinion. This movement is present all over the world (75% of creationists in Saudi Arabia) and even in Europe. Surveys indicate the presence of 8–9% of creationists in Belgium and France (and 55 or 61% of evolutionists respectively) with figures as high as 21% of creationists in Italy[1].

10.2.2. *Platism and the flat Earth theory*

Even more surprisingly, there are groups of people who still defend the Platist theory: the Earth would be a large flat disk, with the North Pole in the center and the huge ice wall of Antarctica all around. The planets and stars would hang from a rotating dome like light bulbs from a ceiling. A growing number of people around the world share this belief in spite of all scientific evidence. There are more than 12 million of them in the United States alone.

The promoters of this theory, in spite of certain divergences concerning the recommended terrestrial model, agree to denounce a generalized conspiracy that is organized by the spatial agencies, GPS manufacturers and

1 https://en.wikipedia.org/wiki/Creationism.

air traffic control services, among others. According to them, the images of the round Earth captured by the space missions are false, and people have never been in space or on the Moon. The scientific data are, however, absolutely irrefutable, and it is difficult to see how a conspiracy of this magnitude could persist in time without being revealed.

In fact, we run out of arguments when we try to counter this heresy. Astronaut Thomas Pesquet, who has made several flights in the European space station, including one of more than six months in 2021, declared in an interview "to be tired of all these people denying the evidence". He captures in one good word the only answer that he considers still possible. For him, the only argument that can still be opposed to the platists is the following one quoted from memory: "If the Earth is flat, I don't understand how our capitalist society that commercializes everything has not yet created touristic voyages to visit its edges."

10.2.3. *The rejection of Western medicine based on evidence*

The first two examples concern theoretical concepts whose application to everyday life is relatively unimportant (if we neglect GPS, etc.). The rejection of evidence-based medicine and its applications, on the contrary, has a much more serious impact. There is in our society a non-negligible fraction of the population which presents a greater or lesser distrust of medicine and which resorts, at least in the first instance, to various alternative techniques to treat health problems. These are often pseudo-medicines which are not based on any rational evidence, when they are not simply practiced by unscrupulous swindlers with no medical training whatsoever, healers, faith healers and other charlatans.

This usurpation has been well documented in the book *Les pseudo medicines, un serment d'hypocrites* by Jean Brissonnet (2017)[2]. This subject goes far beyond the scope of this book, and we will not discuss in detail how

---

2 This book is available in paper version and is also distributed as a PDF at the price of €12 in a collection directed by Henri Broch, University of Nice Sophia-Antipolis, published by Horizons chimériques. See: www.book-e-book.com/e-books/144-les-pseudo-medecines-pdf-un-serment-d-hypocrites-9782372460101.html.

science evaluates homeopathy, psychoanalysis, acupuncture, osteopathy and thermalism, except to reaffirm that there is no proof of therapeutic efficacy of these pseudo-medicines. Let us briefly recall, as an example, some notes on homeopathy:

– Homeopathy is based on the principle of similarity that originated in the imagination of Friedrich Hahnemann in 1819, according to which any product capable of producing the symptoms of a disease would also be capable of curing it. This idea could never be confirmed.

– It uses crazy ingredients such as earthworms, cat fleas, processionary caterpillars or duck livers that have been putrefied for a month and have no curative value.

– It uses compounds at astronomical dilutions ranging from one drop in a swimming pool to one drop in a cube whose side would be about 2 million times the distance from the Earth to the Sun (around 150 million km). It is easy to see that at these dilutions there is not a single molecule of the original compound left per liter of water (fortunately!), which is of course confirmed by the elementary concepts of chemistry and physics. Avogadro's number 6.10^{23} represents the number of molecules present in one mole. Beyond a dilution of 10^{-12}, there is statistically not a single molecule left of the initial solution. Now, 10^{-12} represents only the 12CH dilution in homeopathy which uses up to 30CH.

– And, finally it is based on the theory of the memory of water explaining that water would keep a trace of the compounds that have been dissolved in it, which would explain the action of homeopathic compounds at extremely low dilutions (Davenas et al. 1988). This theory has been clearly demonstrated to have no support[3].

It is therefore surprising that this practice is so widespread in the population and, more seriously, that it is taught in some medical schools in France.

3 https://en.wikipedia.org/wiki/Water_memory.

All these pseudo-medical approaches have developed partly for objective reasons such as the lack of empathy of some doctors, and also for bad reasons such as the mistrust in official medicine accused of covering up various forms of conspiracy. It is also often said that the pharmaceutical industry is only interested in profit at the expense of the efficacy and safety of medicines. It is certainly true that errors and scandals have occurred, but they are by no means generalized.

Moreover, all human activities, apart from voluntary work, are aimed at making a profit, and it is therefore not abnormal for the pharmaceutical industry to do the same. However, there are strict safeguards to limit and often prevent excesses in this area.

These so-called alternative medicines are at best as effective as a placebo, i.e. a pseudo-treatment (administration of an inert substance or fictitious surgical intervention) which has no efficacy of its own but which nevertheless has a weak action via mainly psychological and secondarily physiological mechanisms. They are often harmless but risk delaying the implementation of an effective treatment and, as such, potentially put the lives of some patients at risk. Finally, it should be noted that in some countries, including France, homeopathy is partially reimbursed by Social Security to the detriment of other effective treatments. As the comedian Jérémy Ferrari mentions in his latest show concerning health care, the money spent each year on reimbursing inactive homeopathic medicines would pay the salary of 2,000 nurses[4]!

10.3. Social networks, fake news, post-truth and alternative truths

When the Internet and the World Wide Web started to develop and to make accessible an impressive quantity of information, I thought, as surely many others did, that this new tool would be a source of progress for society

4 Jérémy Ferrari, "Anesthésie générale". The sketch dedicated specifically to homeopathy is free on the Internet at the following address: www.youtube.com/watch?v=AUjBT6xHcdE.

by raising the global level of knowledge. Twenty years later, we realize that this hope has largely been disappointed. The Internet and the social networks that have appeared in its wake have indeed given a voice to all people without distinction. Thanks to, or rather because of, Facebook and other social networks, any individual with no particular knowledge now has potentially the same audience as a Nobel Prize winner! We have thus seen an incredible number of experts of all kinds expressing themselves indiscriminately on any subject.

This has led to the revival of a term that had fallen into disuse: ultracrepidarianism, i.e. the behavior that consists of issuing peremptory opinions on subjects for which one has no particular competence. This term is derived from the Latin expression “Sutor, ne supra crepidam” or “Shoemaker, not beyond the shoe” from the following anecdote by Pliny the Elder: a shoemaker present in the studio of a painter pointed out an error in the representation of a sandal. The painter corrected it. In the wake of this, the cobbler made other criticisms, but the painter replied “a shoemaker should not give his opinion beyond the shoe”. This was the word of the year 2021! The philosopher of science Etienne Klein has contributed to make this term fashionable again on the occasion of the Covid-19 pandemic, which everyone seems to understand, model and treat without having any medical, epidemiological, logical or virological skills.

However, if in some fields, it is legitimate to have divergent opinions, it is not the case for all subjects. We can discuss whether Bordeaux wines are better or worse than Burgundy wines and here all opinions are defensible. But in other fields, the discussion is meaningless: there are unquestionable data that settle the question. And even in cases where science does not yet have an explanation for a series of observations, this is not a reason for filling this void with any nonsense not based on any rational model.

Thus, opinions may be expressed on certain subjects, but there is no reason to have different opinions about firmly established facts. Whether the Earth is flat or round is not decided by a vote or a referendum. However, in the last few years, we have seen large amounts of messages appear on social networks challenging established truths. These ideas, which are essentially lies, have been called "alternative truths" and contribute to the idea of the existence of a post-truth. These messages rub shoulders with others that simply peddle false information (fake news) aimed at manipulating public opinion. President Donald Trump has largely contributed to popularizing these ideas and, while claiming that the traditional media were spreading fake news, he himself multiplied the lies on Twitter.

Unfortunately, it has been well documented that fake news spreads six times faster than real news. They are therefore very difficult to counter. False information about science is thus widely shared on the Internet and contributes to discredit this rational approach of reality. This is the case with the false information that is regularly repeated on the Internet concerning the supposed toxicity of vaccines. The anti-vaccine movement has been around for a long time and has launched various ideas that have since been shown to be false (aluminum toxicity, links with various diseases, including autism or other forms of nervous degeneration), but these ideas are surviving despite the denials that are regularly expressed in the media. These anti-vaccine movements have been strongly revived by the Covid-19 epidemic now relying, unfortunately very effectively, on social networks.

Systematic misinformation organizations have also been created to propagate false ideas, often with a commercial aim. Unfortunately, this is not new and has been going on since before the advent of the Internet. Thus, associations financed by the tobacco industry have, for several decades, paid venal researchers to produce false science aiming to demonstrate that nicotine was not addictive and that cigarettes were not responsible for cancers. All this, of course, in order to combat the results of academic research that demonstrated the harmful influence of tobacco and fought to obtain its ban or at least limit its use and prohibit the publicity about it.

In the same way, lobbies financed by the oil industry are constantly trying, through false information, to discredit the idea that current global warming has, at least in part, anthropogenic causes, whereas this conclusion is the consensus of scientific researchers from all disciplines who participate in the elaboration of the reports of the IPCC, the Intergovernmental Panel on

Climate Change, to which some 2,500 researchers have contributed. The aim here is obviously to block or at least slow down the adoption of measures that would limit or abolish the use of fossil energy sources (oil, gas, coal).

11

The Solution(s)

What are the means by which we could fight this misinformation and this lack of love for science and thus counteract its harmful effects on the global functioning of society? Unfortunately, there is no miracle solution, but several paths can be considered in parallel. Most of them require a better education and information of the public.

11.1. Popularizing research results by researchers themselves

First of all, it is essential for researchers to be more regularly involved in the dissemination of their knowledge to the general public. This implies making time available for this task, and also learning to communicate better and to simplify the data to make them understandable to the largest number of people. This difficulty of communication between researchers and policy makers or the general public is in fact one of the main themes exploited in the film *Don't Look Up* by Adam McKay, released in 2021 on Netflix. In this film, two astronomers try to convince governments and the general public that a large comet is going to hit the Earth and that a space mission should be organized to divert it. However, there is widespread denial and the President of the United States campaigns on the theme of "Don't look up".

Communication channels could include public lectures and also the production of small, simple but scientifically correct films for broadcast in traditional media such as television documentaries. It is also essential to communicate via messages in social networks and other parts of the Internet. This communication exists, but it should be amplified and find reasonably playful forms to attract a growing audience.

11.2. Develop critical thinking skills

It is also necessary to better educate the public in critical thinking so that it becomes impossible or at least more difficult to sell false notions that have no scientific basis. As the American linguist Noam Chomsky said, "Real education is about getting people thinking for themselves."

The French biophysicist Henri Broch (Université de Nice Sophia-Antipolis) has popularized for more than 30 years the notion of zetetic or science of critical thinking. It is defined as "the rational study of phenomena presented as paranormal, of pseudosciences, and of strange therapies"[1]. Its aim is therefore to promote a critical skepticism in order to distinguish what science is, and therefore should not be the object of personal opinions, and what belongs to belief. Zetetic is thus interested in particular and in a non-exhaustive way in pseudosciences, pseudo-medicine, parapsychology, UFOs (the so-called "unidentified flying objects") and currently controversial subjects such as platism, pesticides, GMOs (genetically modified organisms), vaccination and climate change. Zetetic is based on the exercise of doubt, the verification of information and its critical evaluation. It also requires a certain amount of humility because it could easily lead to the ultracrepidarianism denounced above. One doesn't become a specialist in a given subject just because you have done two or three Google searches on the Internet.

Numerous media make it possible to nourish this search for the truth. Henri Broch himself created in 2002 a publishing house that publishes books dealing with very diverse subjects through zetetic. This is the e-book house, which publishes two collections entitled "*Zététique*" and "*Une chandelle dans les ténèbres*"[2] that could translate as "a candle in the night". The YouTube channel entitled "*La tronche en biais*" by Thomas Durand and Vled Tapas produces videos of various formats and publishes interviews that demonstrate the cognitive biases of which we are all unknowingly victims[3]. ASTEC (*Association pour la science et la transmission de l'esprit critique*, French Association for Science and the Transmission of the Critical Mind) essentially pursues the same goals through videos, conferences and

1 https://en.wikipedia.org/wiki/Z%C3%A9t%C3%A9tique.

2 www.everybodywiki.com/Éditions_book-e-book.

3 www.youtube.com/user/TroncheEnBiais.

interviews broadcast on the Internet[4]. Corresponding media exist in the English language (e.g. https://www.criticalthinking.org/ or https://www.internetmatters.org/issues/fake-news-and-misinformation-advice-hub/).

In addition, it is now possible to obtain information validated by experts in various fields in the online encyclopedia Wikipedia[5]. An online news journal such as *The Conversation* publishes daily in French and also in English an electronic journal containing news articles written by specialists[6]. It is therefore easy today, thanks to the Internet, to obtain information from scientifically reliable sources, but unfortunately the Internet is also, as already mentioned, a major source of false information and post-truths.

Finally, it would be absolutely critical to better explain to the general public the real nature of science. It is not in the first place a sum of immutable knowledge but rather a study method that allows the causes of various phenomena to be identified in an objective and reproducible way. The conclusions are obtained only relatively slowly by multiple approaches involving groups of independent researchers. They are therefore subject to change as research progresses. In the early stages of studying a new problem, it is likely that the conclusions reached will change rapidly, as the general public has become aware of on several topics related to the Covid-19 pandemic (e.g. the usefulness of hydroxychloroquine). However, the accumulation of the results will gradually refine the conclusions, which will eventually stabilize and become consensus on which to base decisions. This being said, minor modifications or extensions of a theory are always possible, especially when new means of study become available.

11.3. Expanding an understanding of basic statistics to the general public

An important aspect of education that should receive more special attention is the understanding of quantitative data analysis and statistics. A significant amount of rumors and misinformation is based on deliberately biased analysis of scientific results. Education of the general public in basic statistics would clearly help to uncover some of these attempts at

4 www.esprit-critique.org/.

5 https://en.wikipedia.org/wiki/Main_Page.

6 www.theconversation.com/fr.

misinformation. It would be critical, for example, to make people understand that if a health problem develops as a result of taking a drug, this does not necessarily mean that there is an unwanted side effect. A temporal correlation can only have causal value if it occurs more frequently than expected spontaneously. In other words, is this problem more frequent in people who have taken this drug than in a similar population of people who have not taken it? This can only be established by a well-controlled quantitative study. This is what the pharmaceutical industry normally does.

Similarly, the efficacy of a drug or vaccine can only be established by a methodologically correct statistical analysis of comparable populations. Falsified interpretations can easily emerge and are exploited in particular by anti-vaccine movements. The example of Simpson's paradox deserves in this context to be briefly explained as it was reported in more detail in the Internet journal *The Conversation* on November 4, 2021. This paradox perfectly demonstrates the dangers of misinterpreting statistical results.

It is common to find on the Internet messages that claim to demonstrate that vaccines against Covid-19 are not effective or even harmful. In hospital data collected in England between June and September 2021, it is noted that for the entire population the mortality rate was about half as high among the unvaccinated compared to vaccinated or partially vaccinated people. This is unexpected, to say the least. However, if we split the data according to the age of the patients, then we can see what is the origin of this counter-intuitive result. The schematic example shown in Figure 11.1 sheds light on this paradox perfectly.

Let us imagine a population of 100 people, 50 aged below 50 and 50 aged above. As seen in the real population, few people are vaccinated in the younger age group (5/50), but this number is much higher in the over 50 age group (45/50).

Young and older populations are therefore very different in terms of vaccination status and also in terms of risk of death. Young people with low immunization status have a low risk of death, whereas those over 50 years of age with high immunization status obviously have a higher baseline mortality rate.

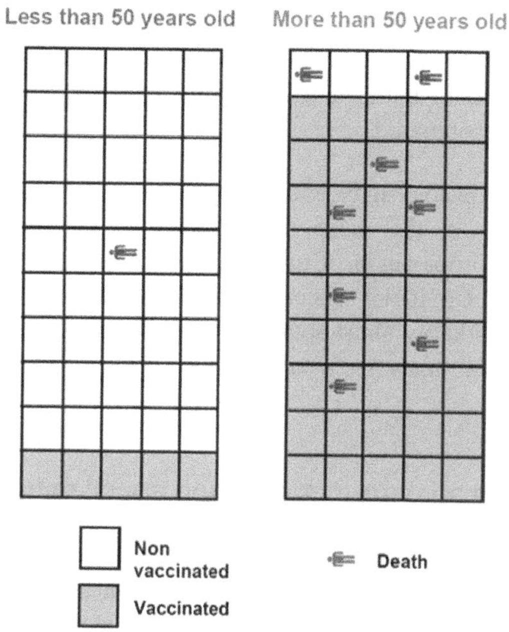

Figure 11.1. *Graphical illustration of Simpson's paradox applied to the analysis of vaccine effectiveness*[7]*. For a color version of this figure, see www.iste.co.uk/balthazart/fiftyyears.zip*

In the youngest age group, there was one death among the unvaccinated (1/45 or 2.2%) and none among the five vaccinated, who are few in number. There is therefore little protection of young people by the vaccine.

At the same time, among the over 50 age group, there were two deaths among the unvaccinated (2/5, i.e. 40%) but six deaths among the vaccinated (red boxes), who were much more numerous, which gives a death rate of 6/45, i.e. 13.3%. There is therefore also protection by the vaccine here (13.3% < 40%). However, if we consider the entire population, we have three deaths among the unvaccinated (3/50, i.e. 6%) and six deaths among the vaccinated (6/50, i.e. 12%), i.e. an absence of protection and even an aggravation by the vaccine.

7 Drawn on the basis of the published article by Berger, Q., Caravenna, F. (2021). Le paradoxe de Simpson illustré par des données de vaccination contre le Covid-19. *The Conversation* [Online]. Available at: www.theconversation.com/le-paradoxe-de-simpson-illustre-par-des-donnees-de-vaccination-contre-le-covid-19-170159.

This paradox, illustrated here schematically with simple numbers, is confirmed in real life with the numbers of the epidemic. It is therefore critical in the statistical analysis of a phenomenon to ensure the homogeneity of the populations compared.

Erroneous conclusions may eventually emerge if there is an undetected heterogeneity, but in this case the paradox is clearly being exploited knowingly in the propaganda of anti-vaccine movements that are trying to discredit the anti-Covid-19 vaccines, an impressive scientific advance achieved in a remarkably short space of time. Training of the population in the basics of data analysis would surely reduce the impact of these manipulations.

11.4. Controlling misinformation using social networks

Social networks, which allow those who live far away to stay in touch (a blessing, of course), play an increasingly important role in shaping ideas in our society. Unfortunately, they have become very efficient amplification boxes for fake news, conspiracy theories and conspiracy theorists among others.

This problem is further compounded by the fact that algorithms such as those developed by Facebook cause users to see only posts from people they actively follow and agree with. This has the effect of isolating people in "opinion bubbles" and making them believe that the whole world agrees with the ideas they are propagating even though these ideas might in fact be those of a small minority. This also leads them by way of consequence to consider the information distributed by the official media (print, radio and television) as false, which is really the world upside down.

This situation is obviously favorable to the development of conspiracy theories that are expressed very freely on the Internet. Entire websites are devoted to the negation of all progress of science and denounce the supposedly corrupt behavior of researchers, politicians who govern us and CEOs of big companies (see namely in French the avalanche of conspiratorial messages and videos posted by Silvano Trotta)[8]. All this targeted disinformation actually creates clans of fiercely opposed opinions. It has even been said that this aspect of the functioning of social networks is

8 www.plus.wikimonde.com/wiki/Silvano_Trotta.

likely to jeopardize democracy. This may be a somewhat catastrophic position, but considering the way in which former US President Donald Trump exploited social networks to instill in a sizeable fraction of the US population the idea that his re-election was going to be stolen, which ultimately led to a violent invasion of the Capitol on January 6, 2021, we have to wonder at the very least.

Targeted censorship is obviously difficult to implement and could also constitute a danger for the freedom of expression. The most obvious excesses must be denounced and blocked, if possible by the managers of the social networks themselves. Donald Trump was effectively banned from Twitter and Silvano Trotta from YouTube where he posted his videos. In this way, a whole activity of filtering and self-censorship of the networks has been developed. They exclude openly conspiratorial messages and videos, which is obviously welcome. About 30,000 videos of disinformation about Covid-19 have been removed from YouTube and 70,000 accounts associated with the complotist association QAnon have been deactivated by Twitter[9]. This type of measure, however, has only limited effectiveness because the Internet remains free and Trump, like Trotta and all the other conspiracy activists affected by the Internet, will quickly reorganize and use or create alternative communication platforms. We can therefore hope to see the development of a self-censorship of social networks and especially a better control of the splitting algorithms that isolate groups of people from each other. We should also hope that researchers will speak out more often on these networks to counterbalance the disinformation that is present. It is only by combining all these approaches that we will be able to maintain a satisfactory presence of rationality and science in the functioning of our societies.

9 www.lemonde.fr/les-decodeurs/article/2021/03/17/chasses-des-grandes-plates-formes-les-reseaux-complotistes-se-reorganisent_6073460_4355770.html.

Conclusion

This book documents the changes that have affected the functioning of scientific research over the past five decades. The comments presented are directly inspired by my career as a biologist involved in the study of neuroendocrinology and behavior. All opinions presented are personal and are solely my own, but having been around many colleagues, I believe they are fairly widely shared. Many of the remarks made most likely apply to other areas of fundamental research.

Research has become infinitely more complex and uses extremely sophisticated methods. This opens up absolutely extraordinary possibilities and allows the study of phenomena and the performance of experiments that, only 50 years ago, would have been considered science fiction. In parallel, unfortunately, the administration of research and the management of the associated risks handicap the researcher who must, in addition, regularly find funds to finance his or her work and the work of the laboratory. This quest for funds consumes a significant part of the working time.

Nevertheless, I consider myself privileged to have been able to practice this profession especially during a part of history in which techniques developed with lightning speed. The second half of the 20th century was, without any doubt, the golden age of biology and I am delighted to have been able to participate, to the extent of my small means, in its development. I still consider that the career of a scientific researcher is one of the most wonderful jobs there is. It gives the opportunity to discover new territories in the company of very intelligent and motivated people.

Of course, access to this career is not easy. It requires long studies, at least four to five years of university, followed by a doctorate of four to five years, often followed by one or two postdoctoral fellowships of a few years. Even then, getting a permanent position in fundamental research will not be easy. Doing a PhD while working 9 to 5 for five days a week will probably not allow us to develop a sufficient *curriculum vitae*, but will nevertheless open doors to other careers that are not without interest.

The selection is therefore severe and only the most motivated and to some extent the most gifted researchers will manage to find a position in fundamental research. They will also have to look for jobs in their speciality anywhere in the world, but if we put in a lot of effort and are flexible about our location, it is possible to find a position as a professor or permanent researcher in the academic world.

The most motivated students from Liège who have done a PhD or a postdoctoral fellowship in my laboratory have often succeeded in doing so, but often by relocating far from Liège to Phoenix Arizona (Pierre Deviche), Austin Texas (Yvon Delville), Paris (Michael Schumacher), Rennes (Thierry Charlier), Tübingen and Shangaï (Henry Evrard), Tours (Matthieu Keller) or Montpelier (Samuel Caro). Others are still in the academic world, hopefully stabilizing their position. Only two of these previous collaborators were able to stabilize their academic position in Liège (Julie Bakker and Charlotte Cornil).

So I am very happy that I was able to find a permanent position in my *Alma Mater* and that I was able to have a full career in research there. I have never in my life felt like I was working. Every morning I woke up eager to get to my laboratory and continue the studies I had started the day before. I think this is one of the greatest joys in life and I would certainly encourage anyone interested in discovering to do the same. I am, of course, a little disappointed that research has largely lost the aura it had when I started my career. This is very unfortunate because humanity is going to face extremely significant challenges in the coming years and science is probably the best way to deal with them. I sincerely hope that our species will have the wisdom to keep rationality at the center of important decisions without giving in to the mirages of alternative beliefs.

References

Abraham, G.E. (1969). Solid-phase radioimmunoassay of estradiol-17 beta. *J. Clin. Endocrinol. Metab.*, 29(6), 866–870.

Apostolakis, E.M., Ramamurphy, M., Zhou, D., Oñate, S., O'Malley, B.W. (2002). Acute disruption of select steroid receptor coactivators prevents reproductive behavior in rats and unmasks genetic adaptation in knockout mice. *Mol. Endocrinol.*, 16(7), 1511–1523.

Arnold, A.P. (2017). A general theory of sexual differentiation. *J. Neurosci. Res.*, 95(1/2), 291–300.

Bakker, J., Honda, S., Harada, N., Balthazart, J. (2004). Restoration of male sexual behavior by adult exogenous estrogens in male aromatase knockout mice. *Horm. Behav.*, 46(1), 1–10.

Balthazart, J. (2019). New concepts in the study of the sexual differentiation and activation of reproductive behavior, a personal view. *Front. Neuroendocrinol.*, 55, 100785.

Balthazart, J. (2020). How technical progress reshaped behavioral neuroendocrinology during the last 50 years... and some methodological remarks. *Horm. Behav.*, 118, 104–682.

Balthazart, J., Blanpain, C., Bureau, F., Cani, P.D., Chariot, A., Drion, P., Engel, V., Hanson, J., De Kerkove D'exaedre, A., Malgrange, B. et al. (2019). L'expérimentation animale : toujours une nécessité pour la santé animale et humaine [Online]. Available at: https://www.lesoir.be/235617/article/2019-07-09/lexperimentation-animale-toujours-une-necessite-pour-la-sante-animale-et-humaine.

Barfield, R.J. (1969). Activation of copulatory behavior by androgen implanted into the preoptic area of the male fowl. *Horm. Behav.*, 1, 37–52.

Bhargava, A., Arnold, A.P., Bangasser, D.A., Denton, K.M., Gupta, A., Hilliard Krause, L.M., Mayer, E.A., McCarthy, M., Miller, W.L., Raznahan, A. et al. (2021). Considering sex as a biological variable in basic and clinical studies: An endocrine society scientific statement. *Endocrinol. Metab.*, 42(3), 219–258.

Bollen, J., Crandall, D., Junk, D., Ding, Y., Borner, K. (2017). An efficient system to fund science: From proposal review to peer-to-peer distributions. *Scientometrics*, 110(1), 521–528.

Bollen, J., Carpenter, S.R., Lubchenco, J., Scheffer, M. (2019). Rethinking resource allocation in science. *Ecol. Soc.*, 24(3), doi 10.5751/ES-11005-240329.

Boyden, E.S., Zhang, F., Bamberg, E., Nagel, G., Deisseroth, K. (2005). Millisecond-timescale, genetically targeted optical control of neural activity. *Nat. Neurosci.*, 8(9), 1263–1268.

Brainard, J. (2022). Silence greets requests to flag retracted studies. *Science*, 377(6601), 11–12.

Brembs, B., Button, K., Munafo, M. (2013). Deep impact: Unintended consequences of journal rank. *Front. Hum. Neurosci.*, 7, 291.

Brissonnet, J. (2017). *Les pseudo-médecines, un serment d'hypocrites.* Éditions Book-e-book.

Burkett, J.P., Andari, E., Johnson, Z.V., Curry, D.C., De Waal, F.B., Young, L.J. (2016). Oxytocin-dependent consolation behavior in rodents. *Science*, 351(6271), 375–378.

Button, K.S., Ioannidis, J.P., Mokrysz, C., Nosek, B.A., Flint, J., Robinson, E.S., Munafo, M.R. (2013). Power failure: Why small sample size undermines the reliability of neuroscience. *Nat. Rev. Neurosci.*, 14(5), 365–376.

Cabrera, H.A. (1969). A comprehensive evaluation of pregnancy tests. *Am. J. Obstet. Gynecol.*, 103(1), 32–38.

Capecchi, M.R. (1989). Altering the genome by homologous recombination. *Science*, 244(4910), 1288–1292.

Carp, J. (2012). The secret lives of experiments: Methods reporting in the fMRI literature. *Neuroimage*, 63(1), 289–300.

Charlier, T.D., Ball, G.F., Balthazart, J. (2005). Inhibition of steroid receptor coactivator-1 blocks estrogen and androgen action on male sex behavior and associated brain plasticity. *J. Neurosci.*, 25(4), 906–913.

Cheng, M.F. and Balthazart, J. (1982). The role of nest-building activity in the gonadotrophin secretions and the reproductive successes of ring doves (Streptopelia risoria). *J. Comp. Physiol. Psychol.*, 96, 303–324.

Clayton, J.A. (2016). Studying both sexes: A guiding principle for biomedicine. *FASEB J.*, 30(2), 519–524.

Clayton, J.A. and Collins, F.S. (2014). Policy: NIH to balance sex in cell and animal studies. *Nature*, 509(7500), 282–283.

Contopoulos-Ioannidis, D.G., Ntzani, E., Ioannidis, J.P. (2003). Translation of highly promising basic science research into clinical applications. *Am. J. Med.*, 114(6), 477–484.

Davenas, E., Beauvais, F., Amara, J., Oberbaum, M., Robinzon, B., Miadonna, A., Tedeschi, A., Pomeranz, B., Fortner, P., Belon, P. et al. (1988). Human basophil degranulation triggered by very dilute antiserum against IgE. *Nature*, 333(6176), 816–818.

De Vrieze, J. (2017). With this new system, scientists never have to write a grant application again. *Science*. doi: 10.1126/science.aal1055.

De Waal, F.B.M. (2010). *L'Âge de l'empathie : leçons de nature pour une société plus apaisée*. Les liens qui libèrent, Paris.

Deisseroth, K., Feng, G., Majewska, A.K., Miesenbock, G., Ting, A., Schnitzer, M.J. (2006). Next-generation optical technologies for illuminating genetically targeted brain circuits. *J. Neurosci.*, 26(41), 10380–10386.

Diester, C.M., Banks, M.L., Neigh, G.N., Negus, S.S. (2019). Experimental design and analysis for consideration of sex as a biological variable. *Neuropsychopharmacology*, 44(13), 2159–2162.

Doudna, J.A. and Charpentier, E. (2014). Genome editing. The new frontier of genome engineering with CRISPR-Cas9. *Science*, 346(6213), 1258096.

Druker, B.J., Sawyers, C.L., Kantarjian, H., Resta, D.J., Reese, S.F., Ford, J.M., Capdeville, R., Talpaz, M. (2001a). Activity of a specific inhibitor of the BCR-ABL tyrosine kinase in the blast crisis of chronic myeloid leukemia and acute lymphoblastic leukemia with the Philadelphia chromosome. *N. Engl. J. Med.*, 344(14), 1038–1042.

Druker, B.J., Talpaz, M., Resta, D.J., Peng, B., Buchdunger, E., Ford, J.M., Lydon, N.B., Kantarjian, H., Capdeville, R., Ohno-Jones, S. et al. (2001b). Efficacy and safety of a specific inhibitor of the BCR-ABL tyrosine kinase in chronic myeloid leukemia. *N. Engl. J. Med.*, 344(14), 1031–1037.

Fisher, C.R., Graves, K.H., Parlow, A.F., Simpson, E.R. (1998). Characterization of mice deficient in aromatase (ArKO) because of targeted disruption of the cyp19 gene. *Proc. Natl. Acad. Sci. USA*, 95, 6965–6970.

Furuyama, S., Mayes, D.M., Nugent, C.A. (1970). A radioimmunoassay for plasma testosterone. *Steroids*, 16(4), 415–428.

Fuyuno, I. and Cyranoski, D. (2006). Cash for papers: Putting a premium on publication. *Nature*, 441(7095), 792.

Garfield, E. (1955). Citation indexes for science, a new dimension in documentation through association of ideas. *Science*, 122(3159), 108–111.

Gibson, G. and Muse, S.V. (2009). *A Primer of Genome Science*. Sinauer Associates, Sunderland.

Gore, A.C. (2013). Editorial: Antibody validation requirements for articles published in endocrinology. *Endocrinology*, 154(2), 579–580.

Harno, E., Cottrell, E.C., White, A. (2013). Metabolic pitfalls of CNS Cre-based technology. *Cell. Metab.*, 18(1), 21–28.

Heather, J.M. and Chain, B. (2016). The sequence of sequencers: The history of sequencing DNA. *Genomics*, 107(1), 1–8.

Hirsch, J.E. (2005). An index to quantify an individual's scientific research output. *Proc. Natl. Acad. Sci. USA*, 102(46), 16569–16572.

Holdcroft, A. (2007a). Gender bias in research: How does it affect evidence based medicine? *J. R. Soc. Med.*, 100(1), 2–3.

Holdcroft, A. (2007b). Integrating the dimensions of sex and gender into basic life sciences research: Methodologic and ethical issues. *Gend. Med.*, 4, 64–74.

Honda, S., Harada, N., Ito, S., Takagi, Y., Maeda, S. (1998). Disruption of sexual behavior in male aromatase-deficient mice lacking exons 1 and 2 of the cyp19 gene. *Biochem. Biophys. Res. Commun.*, 252, 445–449.

Hopewell, S., Loudon, K., Clarke, M.J., Oxman, A.D., Dickersin, K. (2009). Publication bias in clinical trials due to statistical significance or direction of trial results. *Cochrane Database Syst. Rev.*, 1, 6.

Hull, E.M. and Dominguez, J.M. (2015). Male sexual behavior. In *Knobil and Neill's Physiology or Reproduction*, Plant, T.M. and Zeleznik, A.J. (eds). Academic Press, London.

Jalabert, C., Ma, C., Soma, K.K. (2021). Profiling of systemic and brain steroids in male songbirds: Seasonal changes in neurosteroids. *J. Neuroendocrinol.*, 33(1), 12922.

Jallageas, M. and Attal, J. (1968). Determination by gas phase chromatography of non-conjugated plasma testosterone in ducks, quail, and pigeons. *C.R. Acad. Hebd. Seances. Acad. Sci. D*, 267(3), 341–343.

Johnson, R.T. and Dickersin, K. (2007). Publication bias against negative results from clinical trials: Three of the seven deadly sins. *Nat. Clin. Pract. Neurol.*, 3(11), 590–591.

Karp, N.A. and Reavey, N. (2019). Sex bias in preclinical research and an exploration of how to change the status quo. *Br. J. Pharmacol.*, 176(21), 4107–4118.

Kelley, D.B. and Pfaff, D.W. (1978). Generalizations from comparative studies on neuroanatomical and endocrine mechanisms of sexual behaviour. In *Biological Determinants of Sexual Behaviour*, Hutchison, J.B. (ed.). John Wiley & Sons, New York.

Kerr, N.L. (1998). HARKing: Hypothesizing after the results are known. *Pers. Soc. Psychol. Rev.*, 2(3), 196–217.

Krege, J.H., Hodgin, J.B., Couse, J.F., Enmark, E., Warner, M., Mahler, J.F., Sar, M., Korach, K.S., Gustafsson, J.A., Smithies, O. (1998). Generation and reproductive phenotypes of mice lacking estrogen receptor beta. *Proc. Natl. Acad. Sci. USA*, 95(26), 15677–15682.

Lander, E.S., Linton, L.M., Birren, B., Nusbaum, C., Zody, M.C., Baldwin, J., Devon, K., Dewar, K., Doyle, M., FitzHugh, W. et al. (2001). Initial sequencing and analysis of the human genome. *Nature*, 409(6822), 860–921.

Lazic, S.E. (2010). The problem of pseudoreplication in neuroscientific studies: Is it affecting your analysis? *BMC Neurosci.*, 11, 5.

Liere, P., Akwa, Y., Weill-Engerer, S., Eychenne, B., Pianos, A., Robel, P., Sjovall, J., Schumacher, M., Baulieu, E.E. (2000). Validation of an analytical procedure to measure trace amounts of neurosteroids in brain tissue by gas chromatography-mass spectrometry. *J. Chromatogr. B. Biomed. Sci. Appl.*, 739(2), 301–312.

Lisk, R.D. (1960). Estrogen-sensitive centers in the hypothalamus of the rat. *J. Exp. Zool.*, 145, 197–207.

Lloyd, K.C. (2011). A knockout mouse resource for the biomedical research community. *Ann. N. Y. Acad. Sci.*, 1245, 24–26.

Lubahn, D.B., Moyer, J.S., Golding, T.S., Couse, J.F., Korach, K.S., Smithies, O. (1993). Alteration of reproductive function but not prenatal sexual development after insertional disruption of the mouse estrogen receptor gene. *Proc. Natl. Acad. Sci. USA*, 90, 11162–11166.

McCarthy, M.M. and Nugent, B.M. (2013). Epigenetic contributions to hormonally-mediated sexual differentiation of the brain. *J. Neuroendocrinol.*, 25(11), 1133–1140.

McCarthy, M.M., Brooks, P.J., Pfaus, J.G., Brown, H.E., Flanagan, L.M., Schwartz-Giblin, S., Pfaff, D.W. (1993a). Antisense oligodeoxynucleotides in behavioral neuroscience. *Neuroprotocols*, 2, 67–74.

McCarthy, M.M., Schlenker, E.H., Pfaff, D.W. (1993b). Enduring consequences of neonatal treatment with antisense oligodeoxynucleotides to estrogen receptor messenger ribonucleic acid on sexual differentiation of rat brain. *Endocrinology*, 133, 433–439.

McCarthy, M.M., Nugent, B.M., Lenz, K.M. (2017). Neuroimmunology and neuroepigenetics in the establishment of sex differences in the brain. *Nat. Rev. Neurosci.*, 18(8), 471–484.

Meirmans, S., Butlin, R.K., Charmantier, A., Engelstadter, J., Groot, A.T., King, K.C., Kokko, H., Reid, J.M., Neiman, M. (2019). Science policies: How should science funding be allocated? An evolutionary biologists perspective. *J. Evol. Biol.*, 32(8), 754–768.

Meredith, R.W., Zhang, G., Gilbert, M.T., Jarvis, E.D., Springer, M.S. (2014). Evidence for a single loss of mineralized teeth in the common avian ancestor. *Science*, 346(6215), 1254390.

Michael, R.P. (1962). Estrogen-sensitive neurons and sexual behavior in female cats. *Science*, 136(3513), 322–323.

Molenda, H.A., Griffin, A.L., Auger, A.P., McCarthy, M.M., Tetel, M.J. (2002). Nuclear receptor coactivators modulate hormone-dependent gene expression in brain and female reproductive behavior in rats. *Endocrinology*, 143(2), 436–444.

Morrell, J.I. and Pfaff, D.W. (1978). A neuroendocrine approach to brain function: Localization of sex steroid concentrating cells in vertebrate brains. *Am. Zool.*, 18, 447–460.

Mullis, K., Faloona, F., Scharf, S., Saiki, R., Horn, G., Erlich, H. (1986). Specific enzymatic amplification of DNA in vitro: The polymerase chain reaction. *Cold Spring Harb. Symp. Quant. Biol.*, 51, 263–273.

Munafo, M.R., Nosek, B.A., Bishop, D.V.M., Button, K.S., Chambers, C.D., Percie Du Sert, N., Simonsohn, U., Wagenmakers, E.J., Ware, J.J., Ioannidis, J.P.A. (2017). A manifesto for reproducible science. *Nat. Hum. Behav.*, 1, 21. doi: 0010.1038/s41562-41016-40021.

Naule, L., Marie-Luce, C., Parmentier, C., Martini, M., Albac, C., Trouillet, A.C., Keller, M., Hardin-Pouzet, H., Mhaouty-Kodja, S. (2016). Revisiting the neural role of estrogen receptor beta in male sexual behavior by conditional mutagenesis. *Horm. Behav.*, 80, 1–9.

Nicoll, C.S. (1967). Bio-assay of prolactin. Analysis of the pigeon crop-sac response to local prolactin injection by an objective and quantitative method. *Endocrinology*, 80(4), 641–655.

Paabo, S. (2015). The diverse origins of the human gene pool. *Nat. Rev. Genet.*, 16(6), 313–314.

Pennisi, E. (2022). Upstart DNA sequencers could be a game changer. *Science*, 376(6599), 1257–1258.

Pfaff, D.W. (1968a). Autoradiographic localization of testosterone-3H in the female rat brain and estradiol-3H in the male rat brain. *Experientia*, 24, 958–959.

Pfaff, D.W. (1968b). Uptake of 3H-estradiol by the female rat brain. An autoradiographic study. *Endocrinology*, 82, 1149–1155.

Pfaus, J.G., Jones, S.L., Flanagan-Cato, L.M., Blaustein, J.D. (2015). Female sexual behavior. In *Knobil and Neill's Physiology or Reproduction*, Plant, T.M. and Zeleznik, A.J. (eds). Academic Press, London.

Pfenning, A.R., Hara, E., Whitney, O., Rivas, M.V., Wang, R., Roulhac, P.L., Howard, J.T., Wirthlin, M., Lovell, P.V., Ganapathy, G. et al. (2014). Convergent transcriptional specializations in the brains of humans and song-learning birds. *Science*, 346(6215), 1256846.

Raskin, K., De Gendt, K., Duittoz, A., Liere, P., Verhoeven, G., Tronche, F., Mhaouty-Kodja, S. (2009). Conditional inactivation of androgen receptor gene in the nervous system: Effects on male behavioral and neuroendocrine responses. *J. Neurosci.*, 29(14), 4461–4470.

Rhodes, K.J. and Trimmer, J.S. (2006). Antibodies as valuable neuroscience research tools versus reagents of mass distraction. *J. Neurosci.*, 26(31), 8017–8020.

Rich-Edwards, J.W., Kaiser, U.B., Chen, G.L., Manson, J.E., Goldstein, J.M. (2018). Sex and gender differences research design for basic, clinical, and population studies: Essentials for investigators. *Endocr. Rev.*, 39(4), 424–439.

Romero, L.M. and Wingfield, J.C. (2016). *Tempests, Poxes, Predators and People. Stress in Wild Animals and How they Cope.* Oxford University Press, New York.

Sanger, F., Nicklen, S., Coulson, A.R. (1977). DNA sequencing with chain-terminating inhibitors. *Proc. Natl. Acad. Sci. USA*, 74(12), 5463–5467.

Saper, C.B. (2005). An open letter to our readers on the use of antibodies. *J. Comp. Neurol.*, 493(4), 477–478.

Saper, C.B. and Sawchenko, P.E. (2003). Magic peptides, magic antibodies: Guidelines for appropriate controls for immunohistochemistry. *J. Comp. Neurol.*, 465(2), 161–163.

Sar, M. and Stumpf, W.E. (1972). Cellular localization of androgen in the brain and pituitary after injection of tritiated testosterone. *Experientia*, 28, 1364–1366.

Sastre, P. (2014). *Le sexe des maladies.* Favre, Lausanne.

Sauer, B. and Henderson, N. (1988). Site-specific DNA recombination in mammalian cells by the Cre recombinase of bacteriophage P1. *Proc. Natl. Acad. Sci. USA*, 85(14), 5166–5170.

Shendure, J., Balasubramanian, S., Church, G.M., Gilbert, W., Rogers, J., Schloss, J.A., Waterston, R.H. (2017). DNA sequencing at 40: Past, present and future. *Nature*, 550(7676), 345–353.

Stumpf, W.E. (1970). Estrogen-neurons and estrogen-neuron systems in the periventricular brain. *Am. J. Anat.*, 129, 207–217.

Tannenbaum, J. and Bennett, B.T. (2015). Russell and Burch's 3Rs then and now: The need for clarity in definition and purpose. *J. Am. Assoc. Lab. Anim. Sci.*, 54(2), 120–132.

Venter, J.C., Adams, M.D., Myers, E.W., Li, P.W., Mural, R.J., Sutton, G.G., Smith, H.O., Yandell, M., Evans, C.A., Holt, R.A. et al. (2001). The sequence of the human genome. *Science*, 291(5507), 1304–1351.

Wahlsten, D. (1982). Deficiency of corpus callosum varies with strain and supplier of the mice. *Brain. Res.*, 239(2), 329–347.

Watson, J.D. and Crick, F.H. (1953). Molecular structure of nucleic acids; a structure for deoxyribose nucleic acid. *Nature*, 171(4356), 737–738.

Weale, A.R., Bailey, M., Lear, P.A. (2004). The level of non-citation of articles within a journal as a measure of quality: A comparison to the impact factor. *BMC Med. Res. Methodol.*, 4, 14.

Wingfield, J.C. (2017). Ecophysiological studies of hormone behavior relations in birds: Future challenges in a changing world. In *Hormones, Brain and Behavior*, Pfaff, D.W. and Joëls, M. (eds). Elsevier, Amsterdam.

Yalow, R.S. and Berson, S.A. (1959). Assay of plasma insulin in human subjects by immunological methods. *Nature*, 184, 1648–1649.

Yeh, S., Tsai, M.Y., Xu, Q., Mu, X.M., Lardy, H., Huang, K.E., Lin, H., Yeh, S.D., Altuwaijri, S., Zhou, X. et al. (2002). Generation and characterization of androgen receptor knockout (ARKO) mice: An in vivo model for the study of androgen functions in selective tissues. *Proc. Natl. Acad. Sci. USA*, 99(21), 13498–13503.

Zhang, J., Zhao, J., Jiang, W.J., Shan, X.W., Yang, X.M., Gao, J.G. (2012). Conditional gene manipulation: Creating a new biological era. *J. Zhejiang. Univ. Sci. B.*, 13(7), 511–524.

Zhang, G., Li, C., Li, Q., Li, B., Larkin, D.M., Lee, C., Storz, J.F., Antunes, A., Greenwold, M.J., Meredith, R.W. et al. (2014). Comparative genomics reveals insights into avian genome evolution and adaptation. *Science*, 346(6215), 1311–1320.

Index

E, F

G, H

I, K

M, N

O, P

R, S

T, U

V, W, Y, Z

Other titles from

iSTE

in

Biology and Biomedical Engineering

2022

DAMBRICOURT MALASSÉ Anne
Embryogeny and Phylogeny of the Human Posture 2: A New Glance at the Future of our Species (Comparative Anatomy and Posture of Animal and Human Set – Volume 4)

FOSSÉ Philippe
Structures and Functions of Retroviral RNAs: The Multiple Facets of the Retroviral Genome (Nucleic Acids Set – Volume 1)

2021

DAMBRICOURT MALASSÉ Anne
Embryogeny and Phylogeny of the Human Posture 1: A New Glance at the Future of our Species (Comparative Anatomy and Posture of Animal and Human Set – Volume 3)

GRANDCOLAS Philippe, MAUREL Marie-Christine
Systematics and the Exploration of Life

HADJOUIS Djillali
The Skull of Quadruped and Bipedal Vertebrates: Variations, Abnormalities and Joint Pathologies (Comparative Anatomy and Posture of Animal and Human Set – Volume 2)

HULLÉ Maurice, VERNON Philippe
The Terrestrial Macroinvertebrates of the Sub-Antarctic Îles Kerguelen and Île de la Possession

2020

CAZEAU Cyrille
Foot Surgery Viewed Through the Prism of Comparative Anatomy: From Normal to Useful
(Comparative Anatomy and Posture of Animal and Human Set – Volume 1)

DUJON Bernard, PELLETIER Georges
Trajectories of Genetics

2019

BRAND Gérard
Discovering Odors

BUIS Roger
Biology and Mathematics: History and Challenges

2016

CLERC Maureen, BOUGRAIN Laurent, LOTTE Fabien
Brain-Computer Interfaces 1: Foundations and Methods
Brain-Computer Interfaces 2: Technology and Applications

FURGER Christophe
Live Cell Assays: From Research to Health and Regulatory Applications

2015

CLARYSSE Patrick, FRIBOULET Denis
Multi-modality Cardiac Imaging: Processing and Analysis

2014

CHEZE Laurence
Kinematic Analysis of Human Movement

CLARYSSE Patrick, FRIBOULET Denis
Multi-modality Cardiac Imaging: Processing and Analysis

DAO Tien Tuan, HO BA THO Marie-Christine
Biomechanics of the Musculoskeletal: Modeling of Data Uncertainty and Knowledge

FANET Hervé
Medical Imaging Based on Magnetic Fields and Ultrasounds

FARINAS DEL CERRO Luis, INOUE Katsumi
Logical Modeling of Biological Systems

MIGONNEY Véronique
Biomaterials

TEBBANI Sihem, LOPES Filipa, FILALI Rayen, DUMUR Didier, PAREAU Dominique
CO_2 Biofixation by Microalgae: Modeling, Estimation and Control

Printed and bound by CPI Group (UK) Ltd, Croydon, CR0 4YY
16/09/2026
14963077-0008